"YOURS EVER, FREEMAN"

The Wisdom of Freeman Dyson

I0046198

"YOURS EVER, FREEMAN"

The Wisdom of Freeman Dyson

Dwight E Neuenschwander
Southern Nazarene University, USA

World Scientific

NEW JERSEY • LONDON • SINGAPORE • BEIJING • SHANGHAI • HONG KONG • TAIPEI • CHENNAI • TOKYO

Published by

World Scientific Publishing Co. Pte. Ltd.
5 Toh Tuck Link, Singapore 596224
USA office: 27 Warren Street, Suite 401-402, Hackensack, NJ 07601
UK office: 57 Shelton Street, Covent Garden, London WC2H 9HE

Library of Congress Control Number: 2023016691

British Library Cataloguing-in-Publication Data
A catalogue record for this book is available from the British Library.

"YOURS EVER, FREEMAN"
The Wisdom of Freeman Dyson

Copyright © 2023 by World Scientific Publishing Co. Pte. Ltd.

All rights reserved. This book, or parts thereof, may not be reproduced in any form or by any means, electronic or mechanical, including photocopying, recording or any information storage and retrieval system now known or to be invented, without written permission from the publisher.

For photocopying of material in this volume, please pay a copying fee through the Copyright Clearance Center, Inc., 222 Rosewood Drive, Danvers, MA 01923, USA. In this case permission to photocopy is not required from the publisher.

ISBN 978-981-127-185-4 (hardcover)
ISBN 978-981-127-231-8 (paperback)
ISBN 978-981-127-186-1 (ebook for institutions)
ISBN 978-981-127-187-8 (ebook for individuals)

For any available supplementary material, please visit
https://www.worldscientific.com/worldscibooks/10.1142/13295#t=suppl

"Yours Ever, Freeman"

The Wisdom of Freeman Dyson

Dwight E. Neuenschwander, Editor

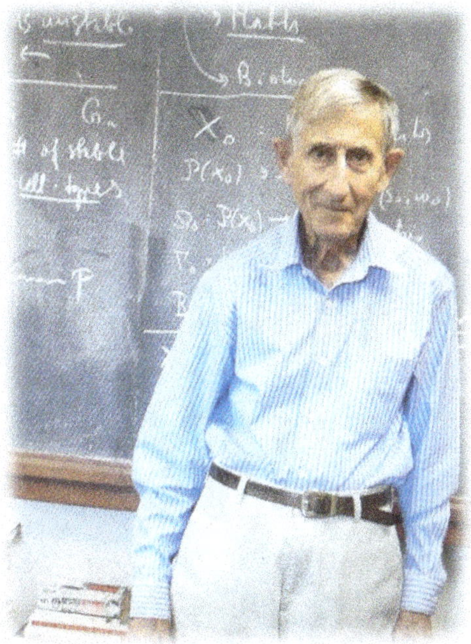

Freeman J. Dyson
Photo courtesy of the Sigma Pi Sigma physics honor society.

"To receive the instruction of wisdom, justice, and judgment, and equity;

to give subtlety to the simple, to the youth knowledge and discretion..." Proverbs 1:2-3

Dedicated to the memory
of Freeman Dyson,
with gratitude,
and appreciation
to his family

Freeman Dyson Maxims

"I never had any difficulty in making friends with people that I disagreed with. Life would be very dull if we could only have friends who agreed with us about everything."

"We must treat our enemies with respect. The ultimate goal must always be, not be to destroy our enemies but to convert them into friends."

"In the end it is how you fight, as well as why you fight, that makes your cause good or bad. A good cause can become bad if you fight for it with means that are indiscriminately murderous."

"To enjoy any sport or any competitive occupation, whether it is football or science, the most important thing is to be a good loser. It is the good losers who make the enterprise enjoyable for everybody."

"The best thing about this planet is that different cultures and different ways of living still survive."

"In the future, as in the past, the great civilizations will be those that give art and science sustained and generous support."

"Technology must be guided and driven by ethics if it is to do more than provide new toys for the rich."

"Science and religion are two windows for looking at the world."

"Maybe I impose my rules on God, but that is not so bad as imposing them on my neighbors."

"Sanity is, in its essence, the ability to live in harmony with nature's laws."

"The principle of maximum diversity says that the laws of nature, and the initial conditions at the beginning of time, are such as to make the universe as interesting as possible."

"These days I spend more time babysitting and less time writing books. You never know which job will turn out to be more important!"

"In some sense, holding babies is an act of worship."

Contents

Preface

The life of Freeman J. Dyson (December 15, 1923–February 28, 2020) will long be celebrated for his many intellectual accomplishments. A representative list of his research contributions include quantum electrodynamics, the Dyson Equations, number theory, random matrices, nuclear engineering, arms control policy, the "Dyson sphere," adaptive optics, astrophysics, cosmological eschatology, the stability of matter, the origin of life, and serious thinking about a long-term vision for humanity expanding across the solar system and beyond. He will also be long remembered as a prolific writer who could effectively communicate to the general public the values and challenges, promises and problems of science and technology. But this book exists to celebrate Freeman Dyson's character as a warm, caring human being—his personal values, his treatment of everyone with respect, his activism for social justice and universal human dignity, his profound sense of community—and his wry sense of humor.

While achieving his professional accomplishments, Freeman Dyson put family first. Beyond his own family, across three decades he also became a wise grandfatherly figure to some 3500 university students as they were trying to figure out who they are while preparing for their life's work. The paths of these students and the path of Professor Dyson intersected in a university course called "Science, Technology, and Society" (STS). Professor Dyson's memoir *Disturbing the Universe* (*DU*) [Dyson, F.J. (1979a)] was and remains the primary textbook in our STS course. From the spring 1993 through the fall of 2019, almost every semester these students corresponded with Professor Dyson, asking questions on a wide and comprehensive range of subjects, from existential concerns about humanity's future to advice for their personal lives.

We were blessed to receive Professor Dyson's insights. His character and wisdom that he shared so freely with us are not ours to keep, but should be shared with everyone. Our first attempt to do this in print came in 2016 with the publication of *Dear Professor Dyson*.[1] In assembling that book its editor had in mind the STS students past, present, and future, making of the book essentially a medley of STS course discussions arranged by topics while highlighting our correspondence with Professor Dyson. More verbose than the correspondence itself, it was a documentary of our journeys as we engaged with Professor Dyson, forming a record for those who took STS before the correspondence began, and for students in years to come. In that role it became a secondary textbook for the STS course.

The present book was motivated by two facts: (1) After *Dear Professor Dyson* was published, the STS classes continued corresponding with him for the rest of his life, and the record should be complete. (2) Unlike *Dear Professor Dyson,* the attempt has been made here to focus on the correspondence itself, setting aside the details of STS classroom discussions and presentations. Here I provide minimal backstories behind the letters, just enough to set their context as required. Beyond that, the correspondence speaks for itself.

Although the correspondence was organized by topic in *Dear Professor Dyson,* here it is presented chronologically. Therefore, the questions presented to Professor Dyson in each letter jump from one topic to another, because the questions were composed near a semester's end when we could survey the entire term's colorful bazaar of subjects. Because the letters often refer to passages in *Disturbing the Universe,* I assume the reader has read it or has a copy readily available.

Although Professor Dyson often describes specific events and individuals in *Disturbing the Universe,* the issues and principles raised by his memoir are timeless. Similarly, even though some of the questions the students put to Professor Dyson referred to contemporary events, his responses are again timeless.

Through our correspondence with Professor Dyson we can see, in his own words and through his recurring themes, his personal values and perspectives on a wide variety of subjects, from the profound to the mundane. His responses to our questions were not always what we expected, and sometimes he disagreed with our statements. In such cases he showed us broader perspectives than we had imagined, and in so doing gave us some education.

In his letters we see Professor Dyson's humility and wisdom that come from a long life well lived. His commitment to social justice and human dignity, his grace and generosity, his self-effacing nature emerge through his letters. We are privileged to see Freeman Dyson, not merely as a legendary remote hero on a pedestal, but as an authentic human being. What an excellent role model human being he is, for us and for all generations!

Dwight E. Neuenschwander, Editor

Note: Quotations of Professor Dyson, whether the passage comes from his letters or published works, appear in italics.

1 In Community

"It is when you give of yourself that you truly give..."
—Kahlil Gibran, *The Prophet*

January 1991: a professor from a small university in Oklahoma attends the Winter Meeting of the American Association of Physics Teachers in San Antonio, Texas. The annual Oersted Medal, the society's highest honor that recognizes a lifetime of significant contributions to physics teaching, is about to be presented to this year's recipient.

The 1991 Oersted Medal recipient is Freeman J. Dyson. Professor Dyson's acceptance speech, "To Teach or Not to Teach,"[1] is to be delivered in the plenary awards session. After one has heard several Oersted Medal acceptance speeches over the years, one comes to expect the distinguished recipient to swoop in on the great day, pose for the customary grip-and-grin photo, deliver a brilliant speech—then disappear after the awards session. But on his day, after Professor Dyson poses for the obligatory photo and delivers a profound speech, he does not disappear. He remains throughout the rest of the meeting, attending parallel sessions, sitting in the audience with colleagues from high schools and universities—another physics teacher in community with the rest of us.

A few days later I am back home with my students in the STS class. Having witnessed Professor Dyson's approachability at the AAPT meeting, I ask the class if they would like to correspond with him. They give a resounding Yes. Thus begins a wonderful, enlightening journey. Looking back over this educational journey of joy and fulfillment, we humbly offer this book as a tribute to Freeman Dyson.

When I joined the faculty of Southern Nazarene University in the summer of 1986, a few weeks before the opening of the fall semester, the General Education director asked me if I would be interested in teaching this new STS course, a course required for all students earning a Bachelor's degree from SNU. I agreed without hesitation, thinking even then how *Disturbing the Universe* would be the perfect textbook. I encountered *DU* when it first appeared in 1979. It grabbed the attention of our little enclave of graduate students who were studying the Dyson equations.[2]

Not only does Professor Dyson's memoir describe events, personalities, technologies, and visions that would provide excellent springboards for STS readings and class discussions, but the book also spoke deeply to me on personal levels. At the level of shared experiences, Professor Dyson's love of literature resonates with this literature appreciator. In addition, in the mid-1950s he spent some time in the vast, silent Great Basin desert, which I also did in the mid-1970s, not far from where he had listened to the sacred silence:

> *It is a soul-shattering silence. You hold your breath and hear absolutely nothing. No rustling of leaves in the wind, no rumbling of distant traffic, no chattering of birds or insects or children. You are alone with God in that silence.* [*DU* p. 128]

Both of us found the Great Basin silence to be a sanctuary.[3]

At the level of contemplating one's place in the universe, the last two chapters of *Disturbing the Universe* hit personally close to home. As one who was raised in a parsonage and immersed in a culture where faith dominates, and as one who was also drawn into science where persuasion requires evidence, the tension alleged by some between science and religion often put me in the bewildering position of striving to build a life based on evidence-based reasoning while, simultaneously, respecting the faith and traditions of parents and grandparents. The last two chapters of *DU* articulated a workable approach to making peace with this tension. They expressed what I felt but had been unable to crystallize into words. So when the General Education director asked me to consider teaching STS at this denomination-sponsored university, I knew that many of the students would come from backgrounds similar to mine. They, too, would be struggling with similar intellectual and emotional dilemmas. Professor Dyson could empathize with us. He had something important to say to fellow travelers, and he articulated it clearly.

These are among the reasons that *Disturbing the Universe* has from the outset been our primary textbook in STS. Through it we have gained a wise mentor and friend.

2 Walking with Grandfather

"The Lakota consider fortitude, generosity, bravery, and wisdom to be the four greatest virtues.... wisdom is not only the greatest of the four greatest, it is also the most difficult to achieve... One has to live a long life to gain wisdom, and it is regarded as life's gift by some who finally achieve it. It is, many realize, a gift they cannot keep to themselves. It must be given back to life,"

—Joseph M. Marshall III, *Walking with Grandfather*[1]

In the spring 1993 semester, the thirty STS students greeted with enthusiasm the suggestion that we correspond with Professor Dyson. Questions for Professor Dyson were solicited, a letter was written, signed by the entire class, and mailed to Princeton.

STS to FD:

April 6, 1993
Professor Freeman J. Dyson
Institute for Advanced Study
Princeton, NJ 08540

Dear Professor Dyson,

We are using your elegant book *Disturbing the Universe* as our textbook in our "Science, Technology, and Society" course at Southern Nazarene University. STS is a general education requirement for all degree-seeking students at SNU. Taken in the junior or senior year, it is a cross-disciplinary view of issues and relations between science, technology, society, and the individual. *Disturbing the Universe* is a perfect text for such a course.... It is a rich book, profound yet accessible to all readers, and a joy to read. For additional background, interpretations, and for following interesting tangents, we occasionally supplement the text with passages from other books and journals, and with films.

> My students and I thought that, as an author, you might enjoy some feedback from your readers. Enclosed you will find from each student a paper which contains one comment and one question. If you could find the time to respond to one or perhaps two of them, we would be very honored. We can appreciate that your calendar is most demanding, so it is with some hesitation that we ask even this much of you. Our request merely reflects how much we have enjoyed and learned from you through *Disturbing the Universe*....
>
> As a member of the audience who heard you present "To Teach or Not to Teach" at the January 1991 AAPT meeting, and as a physicist who is familiar with some of your technical work, it is with a special sense of joy that I have, through your book, been able to share with these wonderful students "the six faces of science,"[2] and introduce to them "the magic city and see in it a mirror image of the big world." [*DU* p. 5] On behalf of all the participants in our course, I thank you for your significant contributions to our lives.
>
> P.S. If you drove through Oklahoma City with Richard Feynman enroute to Albuquerque, [*DU*, Ch. 6] and if you followed Route 66, then you drove right past our campus.[3]

Each of the thirty students submitted a note with one question and everyone signed the letter. We mailed this packet with the awareness that a famous scientist and author might have no time for us, that he had probably never heard of our little university in the Oklahoma prairie. But we hoped for the best, for 'tis better to try and fail rather than failing to try. We clung to the encouraging image of Professor Dyson engaging with all those teachers at the AAPT meeting.

To our delight, within a week we received a two-page letter typed on letterhead of the Institute for Advanced Study, School of Natural Sciences, Princeton, New Jersey. Instead of responding to one or two of our questions, Professor Dyson addressed six of them. We were astonished at the quick turn-around time—his response was written merely three days after our surface-mail letter was dropped into an Oklahoma post office for its 1500-mile trip to New Jersey in those pre-email days. At the next class meeting Professor Dyson's reply was read aloud as the students followed along with photocopies. Although for logistical convenience it was addressed to the STS professor, Professor Dyson's uplifting letter was written to the entire class:

FD to STS:

> *9 April 1993*
>
> *Dear Professor Neuenschwander,*
>
> *The best reward for writing books is to receive a letter like yours. Warmest thanks to you and to your students for your friendly response. Your remarks and your questions uplift my old grandfather spirit. I am sorry I cannot reply at length to each of you. Instead I will give you some brief answers to a few of the questions and send some of my recent talks that touch on the same questions.*[4] *By good luck I just wrote a new preface for the Chinese translation of "Disturbing". Also I will be giving a convocation address at Texas Christian University in Fort Worth next week. Please excuse the repetitions from one talk to another. And please feel free to make copies of any of this stuff and distribute them to anyone who wants them.*

Professor Dyson began each answer by repeating our question back to us. The first question refers to a passage in Chapter 21 of *Disturbing the Universe* where he had written, "In everything we undertake, either on earth or in the sky, we have a choice of two styles, which I call the gray and the green...Physics is gray, biology is green... Factories are gray, gardens are green....Self-reproducing machines are gray, trees and children are green..." [*DU* p. 227] To a student's question asking for elaboration about the distinction between gray and green technologies Professor Dyson responded,

> *1. What is the meaning of gray and green? The distinction here is not between god-made and man-made. Both gray and green may be god-made or man-made. The distinction is between dead and living materials. Gray is made of metal and glass and steam and electricity. Green is made of leaves and roots and cells and enzymes and genes and bugs and brains. Up to now, gray technology has*

mostly been done by engineers, green technology by farmers. In the future the roles of engineers and farmers will become blurred.

In *DU* Professor Dyson described a novel method of storing solar energy as heat—"large ponds enclosed by dikes and covered with transparent plastic air mattresses, so that the water is heated by sunlight and insulated against cooling winds and evaporation...Its energy can be used for domestic heating or converted into electricity..." [*DU* p. 228] Because these insulated ponds could also retain ice for cooling in summer they were called "ice ponds."[5] A student wondered if ice ponds were built at the Institute for Advanced Study.

2. What happened to the ice-ponds? The scheme to use ice-ponds here in Princeton never materialized. The problem with ice-ponds is that they need a lot of attention. They have not yet been packaged so that you can install them and then forget about them. They are being used only by people who enjoy tinkering and taking care of them. I have visited the Kutter Cheese Factory in New York State, which uses ice-ponds successfully and saves a lot of money which would otherwise be spent on electricity for refrigeration. Three facts make this project successful. (a) Mr. Kutter loves his ice-ponds and does not grudge the time he spends messing around with them. (b) The cheese factory has a predictable demand for refrigeration all the year round. (c) The factory is out in the woods and does not need to look elegant. None of these three facts would be true for an average home-owner or for our Institute housing project. Until the ice-ponds can be packaged so as to be neat-looking and easy to maintain, it will not be sold and used by home-owners on a large scale.

3. What piece of technology would I remove if I had the power? My answer to this is nuclear fission technology, assuming that both the bombs and the power-stations would disappear together. Although I do not consider nuclear power-stations evil, it is clear that the benefits of nuclear power-stations are outweighed by the evil of nuclear bombs. There are other benefits of nuclear technology, such as the medical use of radio-isotopes for diagnosis and treatment of

> *many diseases. But on balance, I would be happy to get rid of the power-stations and the medical uses of isotopes if we could get rid of the bombs too.*

We found this response to be very revealing of Professor Dyson's character. He helped design the Triga nuclear reactor, a small reactor used to make isotopes for medical use. The Triga is one of the most commercially successful reactors in the industry. But Professor Dyson would gladly get rid of all nuclear technology—including whatever royalties he might be entitled to from the Triga—in order to rid the world of nuclear weapons.[6]

As a visionary for space exploration, Professor Dyson imagined the human race moving across the solar system and beyond. Drafting foundations on which to build such dreams, he made a career of conducting serious real-world feasibility studies for human engagement in deep space colonization. Professor Dyson writes in *DU* that his "third home" is the long-term future. [*DU* p. 192] A fourth question asked for timescales about people living in space.

> *4. When and how will people be living in space? This is discussed at length in the Tokyo talk.*[4] *The answer depends on what time-scale you are thinking about. If you are thinking about ten or a hundred years, then space-settlements will be unimportant if they exist at all. If you are thinking about a thousand years, then space-settlements will probably be all over the solar system. If you are thinking about ten thousand years or longer, then life will be fully adapted to space and will be spreading over the universe. This is my guess. Of course I may be totally wrong. The purpose of speculating about the future is not to make predictions but to suggest possibilities and broaden our horizons.*

In our first letter of April 1993 several members of the class asked Professor Dyson a range of questions about science and religion, which he compressed into one:

> *5. What effect does science have on my religion? I am not an orthodox Christian, but I am loosely attached to Christianity and*

value the serenity and community that the Church provides. There is no conflict at all between my science and my religion. I consider science and religion to be two windows through which we look out at the world. Neither window by itself gives a complete view. The windows are different but the world outside is the same. Our job as scientists is to explore as much as we can through one window while recognizing that this window gives only a one-sided view. In the talk with the title "Science and Religion,"[7] *I discuss this question a little more deeply. That talk was also expanded into a book, "Infinite in All Directions."*[8]

This reply has echoed through our classes across the years. Professor Dyson's "two windows" metaphor solves a problem for so many of our students who come from religiously conservative homes where science is often treated with suspicion. The two windows motif has consistently found expression in class discussions and student essays, and even in externally-supported seminars.[9] In environments where science and religion have been seen by some as being in irreconcilable conflict, Professor Dyson's two windows metaphor comes as a refreshing and redeeming picture of complementarity.

6. Is there a chance that I might come to visit SNU? Thank you for the invitation, but I have been spreading myself too thin recently, and I decided I cannot take on any more commitments for the next year. I would love to come and talk to each of you face-to-face. But I have a job to do here in Princeton and I am already spending too much time traveling. I am sorry I have to say no.

Fair enough, but had we never asked we would have always wondered if his visit *might* have happened. Grandfather Dyson's friendly closing comments made us feel like we had known him for years:

My apologies to all those whose questions I did not answer. Now it is past midnight and time for me to go to bed. With thanks and good wishes to you all,

Yours ever,
Freeman Dyson

But there was more. Despite the lateness of the hour, on the first page of the letter he added, perhaps as an afterthought, a hand-written note that started in the top margin and continued to the bottom margin. The two halves of the note were connected by a long arrow (Fig. 2.1):

> *P.S. I am glad to see that several of you are studying nursing. The brightest of my daughters is a nurse. It is a splendid profession and gives you the flexibility you need if you want to raise a family. This daughter is also interested in souls as well as bodies and she is a part-time seminary student, intending eventually to be a Presbyterian minister as well as a nurse.*

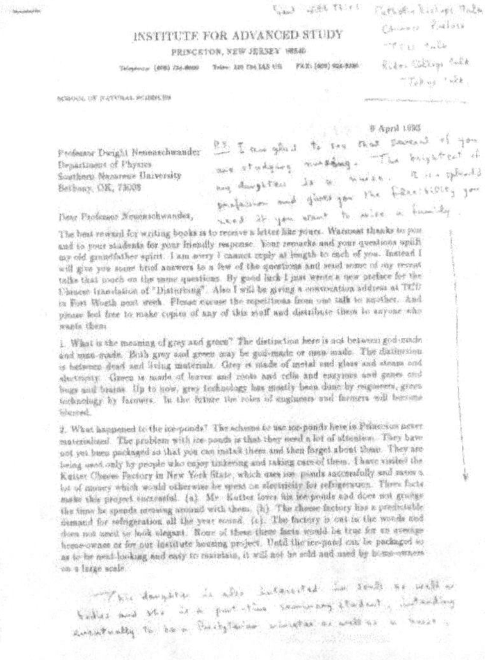

INSTITUTE FOR ADVANCED STUDY
PRINCETON, NEW JERSEY 08540

SCHOOL OF NATURAL SCIENCES

8 April 1993

Professor Dwight Neuenschwander
Department of Physics
Southern Nazarene University
Bethany, OK, 73008

Dear Professor Neuenschwander,

The best reward for writing books is to receive a letter like yours. Warmest thanks to you and to your students for your friendly response. Your remarks and your questions uplift my old grandfather spirit. I am sorry I cannot reply at length to each of you. Instead I will give you some brief answers to a few of the questions and send some of my recent talks that touch on the same questions. By good luck I just wrote a new preface for the Chinese translation of "Disturbing". Also I will be giving a convocation address at TCU in Fort Worth next week. Please excuse the repetitions from one talk to another. And please feel free to make copies of any of this stuff and distribute them to anyone who wants them.

1. What is the meaning of grey and green? The distinction here is not between god-made and man-made. Both grey and green may be god-made or man-made. The distinction is between dead and living materials. Grey is made of metal and glass and steam and electricity. Green is made of leaves and roots and cells and enzymes and genes and bugs and brains. Up to now, grey technology has mostly been done by engineers, green technology by farmers. In the future the roles of engineers and farmers will become blurred.

2. What happened to the ice-ponds? The scheme to use ice-ponds here in Princeton never materialized. The problem with ice-ponds is that they need a lot of attention. They have not yet been packaged so that you can install them and then forget about them. They are being used only by people who enjoy tinkering and taking care of them. I have visited the Kutter Cheese Factory in New York State, which uses ice-ponds successfully and saves a lot of money which would otherwise be spent on electricity for refrigeration. Three facts make this project successful. (a). Mr. Kutter loves his ice-ponds and does not grudge the time he spends messing around with them. (b). The cheese factory has a predictable demand for refrigeration all the year round. (c). The factory is out in the woods and does not need to look elegant. None of these three facts would be true for an average home-owner or for our Institute housing project. Until the ice-pond can be packaged so as to be neat-looking and easy to maintain, it will not be sold and used by home-owners on a large scale.

Fig. 2.1. Page 1 of Professor Dyson's first letter to our STS class.

We don't know if Professor Dyson's children ever saw this footnote, but the other daughters (Fig. 2.2) include a radiologist, a veterinarian, a cardiologist, and a venture capitalist certified as a cosmonaut.

Fig. 2.2. L to R: Freeman, Esther (digital technology consultant and cosmonaut), Dorothy (veterinarian), Emily (cardiologist), Mia (nurse, pastor), Becca (radiologist), Imme (marathon runner). Photo courtesy of the Dyson family.

One is reminded of Garrison Keillor's description of Lake Wobegon, "where all the children are above average."[10] Among the Dyson offspring, that statistical anomaly is true.

Professor Dyson's son is an outdoorsman, a historian of technology, and a superb craftsman who builds by hand beautiful *baidarkas* (Fig. 2.3), canoes of Aleutian design which he constructs with modern materials.[11]

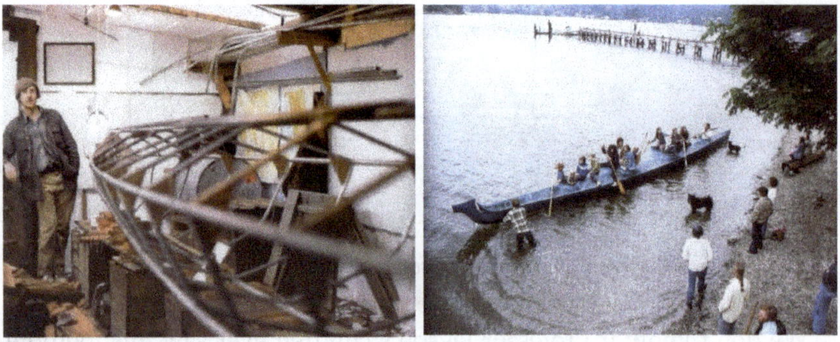

Fig. 2.3. (Left) George Dyson, craftsman and historian of technology. (Right) The "new six-seater kayak" that Professor Dyson mentions on p. 239 of DU. Photos courtesy of George Dyson; see Dyson, G.B., 1997b.

One of his baidarkas is the gorgeous blue "six-seater," the largest baidarka ever built, that Professor Dyson mentions in the "Back to Earth" chapter of *DU*. [p. 239]

As the semester neared its end a thank-you note from the class followed:

STS to FD:

3 May 1993

Dear Professor Dyson,

We, the class of Professor N., felt privileged that a renowned physicist and mathematician such as yourself would take the time as you did to answer our letter of questions.... Please accept our apologies for any lost sleep on our part that we may have caused you!

We were all very delighted to receive not only your answers to our questions, but also the copies of some of your recent lectures and writings...

On behalf of the nursing students, thank you for sharing with us about your daughter being a nurse and a Presbyterian minister. It is important to understand that there is more to a person than just the physical body. We send our best wishes and regards to her as she pursues her nursing and preaching ministries.

Through your book and correspondence, we have seen a more personal side of science and the scientist. Please accept our deepest appreciation and thanks. May God bless you and your family.

Looking to the Future,
STS Class of '93, Spring Semester

We received another reply, a hand-written note on IAS stationary, along with a special announcement:

FD to STS:

May 18 1993

Dear Professor Neuenschwander

Thank you very much for your warm letter and thanks even more to the students for theirs.....

Please give the students my greetings and tell them that I am busy being a grandfather since our daughter organized a very special party for Mothers' Day.

Yours ever
Freeman Dyson

Unofficial Birth Announcement
Donald Dyson Reid, 4 pounds 14 ounces
George Freeman Reid, 5 pounds 4 ounces
made their entrance six weeks early on May 9, 1993.
Dorothy and both kids are doing fine.

When this wonderful note and birth announcement arrived, summer school was underway. But the summer STS class carried the ball. To celebrate the new Dyson grandbabies, we sent Professor Dyson a batch of Father's Day cards and notes of congratulations.

Cautious about wearing out our welcome, our next communication with Professor Dyson did not occur until early 1995. During the winter "mini-term" between semesters, students may take a three-credit-hour course in two weeks, meeting five days per week for four hours per day—a crowded schedule but one that works fine for reading- and discussion-based classes. Because of the limited time, the winter 1995 mini-term students decided to send Professor Dyson a greeting instead of questions. We bought a giant greeting card featuring a *Water Lilies* painting by Claude Monet. Each class member signed the card and wrote a little note to Professor Dyson.

STS to FD:

15 February 1995

Dear Professor Dyson:

...The enclosed card is signed by the students who took the course during the January "mini-term".... I thought you might enjoy hearing these students' comments.

I have found in all semesters (including mini-terms) that the students genuinely like *Disturbing the Universe,* and appreciate how it opens their minds and gets them to think. They say it transforms the way they think about science, since most of them come into the class with images of the usual stereotypes. They say *Disturbing the Universe* is more approachable than a regular text...

On behalf of the several hundred students at SNU [since 1986] who have examined important issues through *Disturbing the Universe*, I thank you for your important contribution to their lives—and to mine.

Warm regards, DN

Again, to our delight he replied with another note on IAS stationary (Fig. 2.4). Although Professor Dyson's reply arrived after the mini-term session was over, I relayed his message to all the mini-term students:

FD to DN and STS:

February 21, 1995

Dear Dr. Neuenschwander

What a delight to hear from you again, and this time with such a heart-warming collection of accolades from your students! Please thank the students, and yourself, for the beautiful Monet lily-pond and the messages inside it. It means a great deal to me to receive a response like this from a new generation of young people. I wish every one of the students as challenging and rewarding a life as I have been blessed with.

Now I am a busy grandfather with our three little grandsons living here in Princeton (ages 3, 1, 1). Lucky again! On Sunday we took all three of them to church and they loved it. As I grow older,

I spend more time baby-sitting and less time writing books. You never know which job will turn out to be more important!

Please keep in touch!

Yours ever, Freeman Dyson

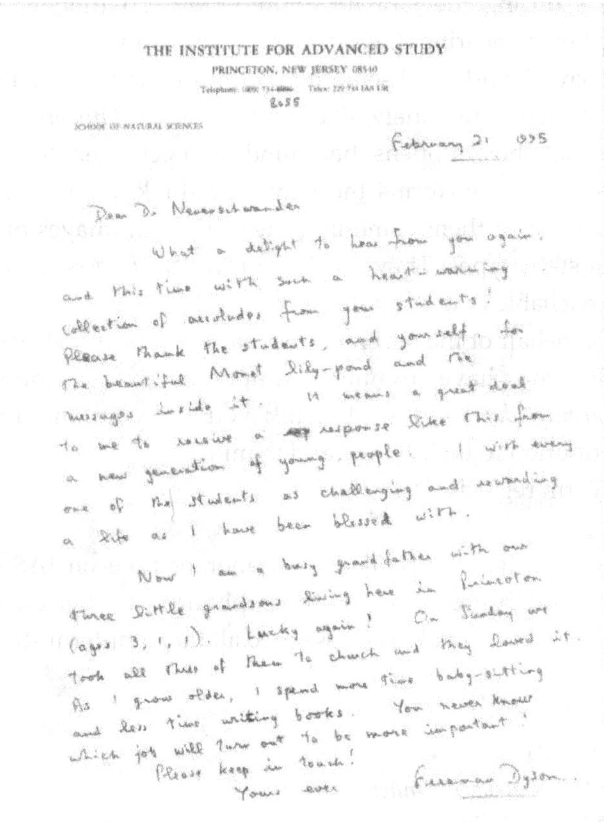
THE INSTITUTE FOR ADVANCED STUDY
PRINCETON, NEW JERSEY 08540

SCHOOL OF NATURAL SCIENCES

February 21 1995

Dear Dr. Neuenschwander

What a delight to hear from you again, and this time with such a heart-warming collection of anecdotes from your students! Please thank the students, and yourself, for the beautiful Monet lily-pond and the messages inside it. It means a great deal to me to receive a response like this from a new generation of young people. I wish every one of the students as challenging and rewarding a life as I have been blessed with.

Now I am a busy grandfather with our three little grandsons living here in Princeton (ages 3, 1, 1). Lucky again! On Sunday we took all three of them to church and they loved it. As I grow older, I spend more time baby-sitting and less time writing books. You never know which job will turn out to be more important!

Please keep in touch!

Yours ever

Freeman Dyson

Fig. 2.4. Writing books and babysitting: "You never know which job will turn out to be more important!"

Professor Dyson's observation about never knowing which job will turn out to be more important offers a guiding principle for anyone trying to balance the demands of profession with the needs of family. When pushing an important deadline, and a tiny person comes into the office and says "Grandpa, will you come play with me?" Professor Dyson's observation springs immediately to mind. Children grow up incredibly fast.

We quickly came to appreciate that, to Freeman Dyson, family comes first.

STS to FD:

4 April 1995

Dear Professor Dyson,

Congratulations on your wonderful and relatively new profession of being a grandfather to three young grandsons…You said in your letter of Feb. 21, "I spend more time baby-sitting and less time writing books. You never know which job will turn out to be the more important!" This is similar to your remark to the survivor of the capsized boat [who was rescued by George, [*DU* p. 243]], "But it seems to me now the best thing I ever did in Princeton was to raise that boy." We find it significant that the last lines printed in *Disturbing the Universe* tell of your raising five daughters, one son, and one stepdaughter.[12] It is so appropriate and meaningful that the text of *Disturbing the Universe* ends in the dream of God's throne, where we find the smiling infant. This profound yet tender image perfectly expresses that which is too deep to be articulated. It has been the source of some of our best discussions in class…

3 A Stone in Chartres Cathedral

"Of course, there had been masters who had enjoyed general esteem, and been recommended from monastery to monastery, or from bishop to bishop. But, on the whole, people did not think it necessary to preserve the names of these masters for posterity... We do not know the names of the masters who made the sculptures of Chartres, Strasbourg or Naumburg. No doubt they were appreciated in their time, but they gave the honour to the cathedral for which they worked."

—E.H. Gombrich, *A History of Art*[1]

With his friendly invitation to keep in touch, Professor Dyson made us feel welcome to continue corresponding with him. Because he responded to six of the original thirty questions sent in our first letter, we henceforth held the volume of questions per semester to about that number.

As the spring 1995 semester approached its final weeks, we collected another batch of questions to send to Professor Dyson. But our letter was suddenly upstaged by a grim event that crashed down on Oklahoma City and rocked the nation. On April 19, 1995, we experienced a terrorist bombing delivered by a rental truck laden with ammonium nitrate and parked in front the Alfred P. Murrah Federal Building. One hundred and sixty-nine people lost their lives, including nineteen babies and toddlers in the second-floor day care facility. When the bomb exploded we felt the shock wave on our campus seven miles away. Through this experience we could better appreciate what the residents of London, including 16-year-old Freeman Dyson, had to endure during the Blitz. We tried to imagine such destruction and loss of life happening every day for six months. A few days after the Murrah bombing, when our class sent questions to Professor Dyson, we prefaced them an extra message:

STS to FD:

April 25, 1995

Dear Professor Dyson,

Please see attached letter. The students have finished reading *Disturbing the Universe* and are in a mood to discuss.

> Update from downtown Oklahoma City: The city still mourns as the grim work goes on. Even now the rescue workers are approaching the ruins of the day care center, buried under tons of rubble, containing the little ones.... But this is a city filled with grace. Whatever is needed by the victims, families, and rescue workers, when the word is put out the community responds instantly and generously. The cowardly and evil actions of the few are no match for the outpouring of love and grace of the many. "Set like a seal upon thy heart, Love is stronger than Death."[2]

The next day Professor Dyson replied to our questions. But first he responded to our preliminary note about the events in Oklahoma City. He recalled being that sixteen-year-old who experienced the London Blitz:

> *26 April 1995*
> *Dear Dwight Neuenschwander,*
> *Thank you for your moving report from Oklahoma City, and for the questions from the students. What you say about Oklahoma City brings back memories of the London blitz. The spirit of community and strength and brotherly love was exactly as you describe it. We had the additional advantage that there was little hate for the German boys in the sky who were also risking their lives. We knew that we and the boys overhead were all in it together.*

His recollections of the Blitz, alongside our observations of the Murrah bombing aftermath, motivated another letter to Professor Dyson on this topic a few weeks later, in May. Jumping to it ahead of our questions of April 25 and his answers of April 26, let me share that May letter with you here:

STS to FD:

> May 16, 1995
> Dear Professor Dyson,
> ...Thank you for your gracious words about the "spirit of community and strength and brotherly love" that have come

out of the tragedy in Oklahoma City. Your words in *Disturbing the Universe* were so appropriate to the events of the past few weeks.... For instance, "In the end it is how you fight, as much as why you fight, that makes your cause good or bad." [*DU* p. 41] Evidently the bombers of the Murrah Building thought of themselves as waging war on the federal government. To use tactics that include truck-bombing to oblivion 168 innocent people, including 19 pre-school children, reminds us again that the ends do not justify the means....

I don't believe there is a single family in Oklahoma City and the surrounding area that was not personally involved. If you didn't know someone who was lost or injured in the building, you quickly found someone in your second tier of contacts—friend of a friend, relative of a co-worker, etc. Many of the students in the class volunteered for service in various way in the days after the bombing, distributing supplies to rescuers, or counseling, or working in the downtown convention center that was converted into a hotel for the rescuers from other cities. From donating time and supplies, to hastily organizing a substitute birthday party for a small boy whose parents were suddenly called downtown to identify an aunt's body, everyone was directly involved. The way the community pulled together and put their arms around the victims and their families, is the finest redeeming grace in all of this. Indeed, it touched the entire country. Not only did my own children's classes write letters to the rescuers, but children all over the country wrote letters to Oklahoma City children....

The fury of the bombers is no match for such a spirit. Perhaps now those of us who did not experience the Blitz or Dresden or Hiroshima ourselves can better appreciate, at least in some approximate way, what values and principles are tapped by those on the receiving end of such disasters. Some important differences you had to live with in the Blitz was the knowledge that the next bomb might land on your

> house, and that almost everybody had a close friend or relative who was lost....
>
> Warm regards, DEN & STS

As the world sadly knows, since April 1995 we have seen too many of these cowardly attacks on the innocent, done in the name of some cause that, however good it may (or may not) have been originally, is made bad and irrelevant when the perpetrators use "means that are indiscriminately murderous." [*DU* p. 41]

Returning to our April 25 letter, from a list of about a dozen questions the 57 students ranked each one, and the most popular few were sent to Professor Dyson:

STS to FD:

> ...The class would consider it a great honor if you could respond to one or more of these questions.... We know that you are very busy, and thus any response which you might have would be greatly appreciated....
>
> 1. Referring to the last chapter, "Dreams of Earth and Sky," what does the baby represent—what specifically did you have in mind? What attributes of God are you trying to represent?

This question refers to the closing scene in *DU.* In this last dream Professor Dyson has an appointment with God. Two of his daughters come along. Upon arriving in God's throne room, they see a wicker throne, which appears to be empty, at the top of some steps. Maybe God did not expect them to be so punctual. After waiting some time, Freeman ascends the stairs and finds a three-month-old baby smiling at him. He and his daughters take turns carrying the baby, then Freeman gently places him back on his throne and says goodbye. Freeman's questions remained unasked.

Professor Dyson did not specify the questions he was going to ask God, nor did he explain the significance of a smiling infant occupying God's throne. Like great music or literature, through image and mood this scene describes something which words cannot articulate. The reader can place himself or herself in the dream, with his or her own questions. We wondered if Professor Dyson would annotate his dream.

FD to STS:

26 April 1995

Dear STS class,

It was lucky that you send the letter by FAX, as today is the last day I have any time to answer it. Tomorrow we start three full days of astronomy meetings, and then I fly off to Jerusalem for three weeks of lectures at the Hebrew University. So I give you my answers to the questions as best I can, without having time for careful thinking.

Question 1. What does the baby represent? What specifically did you have in mind? What attributes of God are you trying to represent? The main fact concerning the dream is that it was a genuine dream, not a consciously composed story. I had nothing in mind when I dreamed it but the dream itself. I did not think about what the dream was supposed to represent until much later.

I should say that the account of the dream in the book is incomplete in one respect. In the actual dream, there was an intense and overwhelming flood of joy that streamed through me while I was holding the baby. It was not that my questions were answered, but my questions were swept away by the power that streamed through me. Afterwards, when I wrote the concluding pages of the book, I decided to leave out this aspect of the dream for artistic reasons. As a writer, I wanted the book to end with a quiet good-bye, not with a flourish of trumpets. So I toned down the ending. I still think it was better to give it a quiet ending. The ending as written is not untrue, only incomplete.

To come back to the question of what the dream means. In my waking life, holding babies is about as close as I ever come to a personal religion. In some sense, holding babies is an act of worship. When I am holding a baby, it often reminds me of the Bergman film "The Seventh Seal," in which the juggler Jof with his young wife and baby sit snug while the storm rages and the angel of death flies over their heads. So for me every baby carries a message of life and hope and survival. It is natural to think of a baby as a little bit of

> *God. But the dream is not an intellectual exercise. It is rather a mystical exercise, or an unconscious work of art. I am not "trying to represent" anything. You might say, God is representing himself in the only way I can understand.*
>
> *If I try to answer your question by interpreting the dream in intellectual terms, I would say: God is not an end but a beginning, a part of the universe that was only recently born. But that is only an after-thought.*

One does not often hear a distinguished physicist speaking of mystical exercises. But Professor Dyson's experiences and interpretations remind us that there may be more to ultimate reality than can be grasped by our finite minds. We need two windows for a complete view.

STS to FD:

> 2. For what do you wish to be the most remembered? The students have learned that you are a mathematician, physicist, father, grandfather, a figure in public policy affairs....

FD to STS:

> *Question 2. For what do I wish to be the most remembered? I take some pride in doing a number of different things well, and I do not much care which thing will be remembered longest. My children and grandchildren are a source of great pride and joy, but to their grandchildren I will be only a name. My work as a scientist was like putting a few stones into one of the arches during the building of Chartres cathedral. The science will endure as a thing of majesty and beauty long after my personal contribution is forgotten. At the moment I am best known to the public because of the "Dyson Sphere" that appeared on the Star Trek program.*[3] *This is a silly joke, but still I enjoy the fame that brings me closer to the Star Trek generation. I suppose in the long run it will probably be the writing of "Disturbing the Universe" that has the best chance of leaving a personal trace of me in the memory of future generations. But I am quite satisfied if the book speaks to your generation. Whether it lasts longer than that is not important.*

It has been said that there are two kinds of deaths.[4] First comes the inevitable death of the physical body. But the second death—not necessarily inevitable—occurs when one is forgotten. Professor Dyson seems to not fret about either one. His serenity does not depend on what others think, either now or in the time to come. It is enough to have laid a stone, however anonymously, in Chartres Cathedral.

Chartres Cathedral, Chartres, France (D.E.N. photo).

In *DU* Chapter 21, "The Greening of the Galaxy," Professor Dyson muses on the long-term fate of the human race across vast timescales. "The question that will decide our destiny is not whether we shall expand into space. It is: shall we be one species or a million?" [*DU* p. 234] Having read these passages and been introduced to Professor Dyson's 1979 paper "Time Without End: Physics and Biology in an Open Universe,"[5] the students raised the next question.

STS to FD:

3. What do you think will become of our civilization?

FD to STS:

Question 3. What do I think will become of our civilization? I tried to answer this question as best I could in the last chapter of "Infinite in All Directions".[6] *Of course nobody is wise enough to foresee the march of history. Fate will play us all kinds of unexpected tricks, good and bad. But the future is still largely in our own hands.*

History gives us opportunities to make choices, to take advantage of Fate's tricks in one way or another. My firmest belief about the future is that within a few hundred years life will be streaming out from the earth all over the solar system and beyond. Once the expansion of the domain of life is started, we will be powerless to stop it even if we wanted to. If we are wise, we will go along with it and adapt ourselves to whatever opportunities we find. Then our civilization will have chances to grow and diversify in ways we cannot imagine.

FD to STS:

4. What is the most important scientific discovery of the 20th century?

Thank you, Professor Dyson, for all your communications with us, from *Disturbing the Universe,* to your speeches and papers, to your personal correspondence. You have truly communicated with us and have taught us much.

Warm regards,
STS, Spring 1995 semester

FD to STS:

Question 4. What is the most important scientific discovery of the twentieth century? Here I have nothing original to say. To me the most important discovery was the double helix structure of DNA, showing that the basic processes of biology are understandable in terms of ordinary chemistry. This does not mean that life is "reduced" to nothing more than a sequence of chemical reactions.[7] *It means that life is approachable using the methods of science. Life can be studied and sculptured using the tools of chemistry. Compared with this, relativity and quantum mechanics are of minor importance....*

I will be back from Jerusalem in May 25 and look forward to reading any more comments and reactions from the students then. Meanwhile, I wish them a happy final exam and thank them for their questions.

Yours sincerely,
Freeman Dyson

4 Two Windows

"Faith" is a fine invention
When Gentlemen can see—
But Microscopes are prudent
In an Emergency.
—Emily Dickinson

From June 1995 through August 1997 I found myself on a leave of absence from SNU, serving in a management role within the American Institute of Physics in College Park, Maryland. This meant not being able to teach STS for the duration, which reduced the flow of STS-Dyson correspondence. Even so, during this time the flag for the effectiveness of *Disturbing the Universe* as a pedagogical resource was flown in the form of a contributed talk at another AAPT meeting.[1]

On Thursday, October 12, 1995, Professor Dyson was the guest speaker in the weekly colloquium of the George Washington University Department of Physics. That afternoon I slipped out of the office and rode the Metro to GWU. Professor Dyson's title for his colloquium talk was "New Directions in Applied Physics." The abstract was typical Dyson in its clarity and its focus on the next generation of students:

> *The content of this address is primarily a pep-talk directed mainly at undergraduate physics majors, describing a handful of exciting ventures now in progress, in which people trained as physicists are applying their skills to cause revolutions in other sciences, such as medicine, chemistry and ecology. The main message of the talk is that an education in physics is valuable even if the job-market in cosmology or string-theory is less than heartening. You don't have to be a cosmologist or a string-theorist to do exciting stuff!*[2]

Physics majors taking their physics degrees into diverse professions are a community of friends of physics distributed throughout the larger society. The accessibility to diverse professions that comes with a physics degree deserves amplified attention and celebration. Although I had STS reasons to meet Professor Dyson in person, his message that afternoon at GWU resonated beautifully with what we at AIP were trying to do with the Society of Physics Students and the Sigma Pi Sigma Honor Society.[3]

After his talk, in the hallway I introduced myself to Professor Dyson. He was very friendly and glad to see a face behind the STS letters. Following a get-acquainted chat, in the course of our conversation I asked him a question about recent changes in US nuclear weapons policy. It featured a moratorium on nuclear testing, leading to the Stockpile Stewardship program where testing would be done only by computer simulations. Professor Dyson described this development as the high point of the first Bush Administration.

During my waning days of my appointment at AIP I received a package from the Institute for Advanced Study. Inside was the book *Darwin Among the Machines* by Freeman's son George. In his preface George recalls his career on fishing boats, and wintering high in a snug, sturdy tree house equipped with a fireplace and glazed windows. George built this house a hundred feet aloft in a Douglas fir. He recalled drifting off to sleep in this tree house and, living with the trees, wondered whether trees might think—"not thinking as you or I think, but thinking the way trees think, taking a hundred years to form an idea." While keeping watch in a fishing boat during the pre-dawn hours, he muses,

> "When you live within a boat its engine leaves an imprint, deeper than mind, on neural circuits first trained to identify the acoustic signature of the human heart. As I had sometimes drifted off to sleep in the forest canopy, boats passing in the distance, and wondered whether trees might think, so I sat in the engine-room companionway in the small hours of the morning, with the dark, forested islands passing by, and wondered whether engines might have souls."[4]

Soulful engines would emerge as a question in a future STS letter to Professor Dyson, to which he would give an interesting response.

At the start of the spring 1998 semester I was back in the STS classroom. Our discussions included public education, nuclear weapons history and policies, environmental sustainability, genetic manipulation, science and religion. Those interests emerged when our letters to Professor Dyson resumed.

STS to FD:

> 1 May 1998
>
> Dear Professor Dyson,
>
> ...The students in the course have raised the following questions which they would like to ask you, should you have the time to consider them....
>
> 1. From "The Argument from Design" [*DU* Ch. 23]—how would you integrate science and religion?

This question was inspired by a passage from Ian Barbour's Gifford Lectures, later published as *Religion in an Age of Science.*[5] Barbour classifies relationships between science and religion under four headings: Conflict, Independence, Dialogue, and Integration. The first three of these were clear enough, but the last one raised some questions for us locally. Some of our university administrators often spoke of "integrating faith and knowledge." It was never spelled out what this really meant or how far it was intended to go. On one hand, Zoroastrian Astrophysics and Manichean Antenna Design are not distinct disciplines. Competence in astrophysics or antenna design requires no particular religious outlook or the lack of one, and viewing a secular discipline through a religious doctrinal lens will likely result in confusion and disservice to both the secular discipline and the religion. On the other hand, at the level of one's personal peace-making with the our place in the universe, criteria for a responsible, honest position was well articulated by the great Princeton biologist Edwin Grant Conklin. Before becoming a biologist, as a young man Conklin considered the ministry. In an illuminating article published only few weeks after the famous Scopes "monkey trial," Conklin wrote that religion must be reconciled to knowledge, not the other way around.[6] "Integrating faith and knowledge" can be either a noble quest for harmony—or it can be wielded as a bludgeon of ideological conformity—hence our question for Professor Dyson.

FD to STS:

2 May 1998

Dear Dwight and Class,

Thank you for your message and the questions. Since time is short I give you some quick answers. Since I am retired I am busier than ever before. Traveling too much and pontificating wherever I go. Luckily your message arrived during a brief visit home between two trips.

1. The quick answer is, I don't integrate science and religion. Science and religion are two windows that give us different views of the world outside. Both are valid, but we can't look through both windows at the same time. I attach to this message a chapter with the title "The Two Windows" explaining what this means. The chapter belongs to a book "How Large is God?" edited by John Templeton and published a few months ago.

In that volume Professor Dyson emphasizes the "two windows" metaphor again:

> *Religion and Science should not look at each other as two systems of laws which must be forced into exact accord. A better metaphor to describe religion and science today is two windows, looking out on the world in different directions....Why cannot we look through both windows simultaneously? The essence of religion is faith and the essence of science is doubt.*[7]

As it often did, the final chapter of *DU*, "Dreams of Earth and Sky," motivated another question.

STS to FD:

2. Is there a dream of yours that remains unfulfilled?

FD to STS:

2. Many dreams remain unfulfilled. I suppose the most important is the dream of Samuel Gompers, that the United States should become a society of schools, books and leisure instead of a society of jails, guns and greed. For this, see the last chapter of my book "Imagined Worlds."[8]

STS to FD:

3. How far can we go and ethically conduct research in genetic engineering on humans?

FD to STS:

3. Here I recommend the book "Remaking Eden," by my friend Lee Silver who is a professor of biology here in Princeton.[9] *The book is mostly about fertility clinics and the things people will do in order to have babies that they can call their own. Before you can write laws to regulate genetic engineering applied to humans, you must look at it from the point of view of the parents. Obviously the laws should prohibit practices that carry higher risk than natural conception of giving birth to defective babies. But it is not possible to require the informed consent of a baby before it is allowed to be born. In my opinion, the laws should be written slowly, in response to real problems as they arise. It would be a big mistake to write laws in a hurry in response to political or ideological pressures.*

Instead of reacting to political urgency or ideological fears, making policies in thoughtful response to experienced realities would be a good practice for all our elected and appointed leaders!

STS to FD:

4. How would you respond to the following passage from John Steinbeck—"Fear the time when the bombs stop falling while the bombers still live, for every bomb is proof that the spirit has not died."

This passage comes from Chapter 14 of *The Grapes of Wrath*. The impoverished tenant farmers are driven off the farms by the combined pressures of the Great Depression, the Dust Bowl, grasping corporate farming, and mechanization. The farmers and their families pile all they can carry onto their worn-out jalopies and, with high hopes, set out from Oklahoma to the vineyards, orchards, and farms of California. But along the way they are routinely cheated and treated with contempt. When they arrive in the Golden State they are manipulated, terrorized, demonized and abused. Beneath their disillusionment and desperation festers a growing anger. The grapes of wrath begin to ripen. These "Okies," willing to work hard at any job, however menial, simply want to take care of their hungry families and be treated with human respect. But they are met with hostility and brutality, misled and deprived of dignity and justice. The grapes of wrath continue to ripen. But so long as ordinary people do not abandon the fight for social justice, hope still lives. Abandonment of the fight leads not to peace, but to desolate resignation, as Steinbeck observes:

> "This you may say of man—when his theories change and crash,...man reaches, stumbles forward, painfully, mistakenly sometimes. Having stepped forward, he may slip back, but only half a step, never the full step back... If the step were not being taken, if the stumbling-forward ache were not alive, the bombs would not fall, the throats would not be cut. Fear the time when the bombs stop falling while the bombers live—for every bomb is proof that the spirit has not died. And fear the time when the strikes stop while the great [land] owners live—for every little beaten strike is proof that the step is being taken. And this you can know—fear the time when Manself will not suffer and die for a concept..."[10]

Professor Dyson extends the interpretation of this passage from the era of victimized Okies struggling against the threat of abuse and starvation, to our era's struggle against the threat of nuclear annihilation. Fear the time when no one cares enough to push back against nuclear Armageddon.

FD to STS:

> *4. I like the Steinbeck quote. So far as the present situation is concerned, I take it to mean that we should get rid of nuclear weapons as quickly and completely as possible. So long as the bombs exist, there will be people with the power to use them, and with the spirit that is willing to use them in anger.*

STS to FD:

5. After fossil fuels are depleted, will nuclear energy be our primary energy source?
Thank you once again Professor Dyson for touching our lives.
Warm regards, The STS class

FD to STS:

5. I see no reason to rule out nuclear energy as a major source of power after the oil is gone. But it is likely that solar energy will be the primary source, because it is more abundant and distributed more evenly over the earth. Solar energy has many advantages in flexibility and adaptability to local conditions. At the moment solar energy is more expensive than nuclear energy, but in the long run solar energy will probably be cheaper. And solar energy is most abundant in the tropics where most of the people live.

That's the best I can do. Please give my greetings to the students and tell them I feel honored that they read my stuff. Yours ever, Freeman.

Our next letter was sent near the end of the Fall 1998 semester.

STS to FD:

3 December 1998
Dear Professor Dyson,

At the risk of imposing on your past graciousness, we have gathered questions from every student in our 50-person STS class this semester. We narrowed the field to the most recurring few. If you have the time to respond to one or two of them, the class would be most appreciative....

1. Do you foresee boundaries to science and technology? How far can or will science go? What happens when we reach the limit?

FD to STS:

5 December 1998

Dear Dwight,

Thank you for your message. Please give my greetings to your students and tell them I am glad to see their questions even if I don't have time to answer them adequately. So here are my answers....

1. We cannot possibly see from here whether there are any limits to science or to technology. Responding to John Horgan's book, "The End of Science,"[11] *my son said, "How can he know about the end of science when we are still so close to the beginning?" What my son says is true. We are still close to the beginning of science. My own guess is that science of some kind will continue to develop so long as the human species survives. But the science of AD 4000 will probably be as different from our science as our science is from the science of Aristotle. It might be so different that we would no longer call it science. But I cannot imagine how we could "reach the limit." So long as there are people, there will be new things to explore and new questions to answer.*

STS to FD:

2. What motivated you to write *Disturbing the Universe*?

FD to STS:

2. I was invited by the Sloan Foundation to write a scientific autobiography, as one of a series that they were publishing. I was glad to accept the invitation, because I enjoy writing. I had done enough interesting things and known enough interesting people, so I knew I had material to make a book. The most important thing when you are writing a book is to have something to say. I had something to say. Also the Sloan Foundation paid me a substantial advance, which helped to put my daughters through college. Since we have five daughters, college bills were a strong motivation.

The STS students—and university students everywhere—relate to Professor Dyson's latter point!

The next question was even more personal, but Professor Dyson's reply reminds each of us that we owe much to many people.

STS to FD:

3. Which individual had the greatest influence on your life, and why?

FD to STS:

3. No single person had the greatest influence in my life. Different people are important at different times, and you can't put them in order of importance. Probably the three most important were my mother, from age zero to twenty, Richard Feynman, from age twenty to thirty, and my wife, from age thirty to seventy-four. My mother gave me a splendid education, Feynman gave me my chance to do an important piece of science, and my wife gave me a big family, so I was three times lucky.

STS to FD:

4. We might compare your life to a modern-day Benjamin Franklin, a Renaissance man who contributes to almost every important part of the culture. What were the most important lessons you have learned in your life?

FD to STS:

4. I have not learned any big lessons that can be expressed in words. To live well is an art and not a science. My life has been opportunistic, not making big plans but responding to opportunities wherever they arose. Perhaps the main lesson is, always be ready to jump at the next unexpected opportunity. The Sloan Foundation invitation to write a book was a good example. In the last twenty years I have learned more from my children's lives than from my own. They are all good at jumping at unexpected opportunities, and as a result they all have interesting lives.

STS to FD:

5. What do you consider to be science's biggest mistake?

FD to STS:

5. Science's biggest mistake happened in 1939 after nuclear fission was discovered, one year before the beginning of World War Two. The physicists could have organized an international meeting of experts to discuss the problem of nuclear weapons, as the biologists did in 1975 when the sudden discovery of recombinant DNA technology made genetic engineering possible. The biologists agreed on a set of rules to ban dangerous experiments, and the rules have been effective ever since. The physicists could have done something similar in 1939, and there was a good chance that nuclear weapons would never have been built. But the chance was missed. Once World War Two had begun in September 1939, it was too late, because the scientists in different countries could no longer communicate.[12]

STS to FD:

6. What comments do you have on current developments in genetic manipulation? Have we already gone too far to turn back?

FD to STS:

6. I am delighted with the progress in genetic manipulation, because it can help enormously to feed hungry people (when applied to crop-plants) and to cure diseases (when applied to viruses and humans). I would certainly not wish to turn back, even if it were possible. It is impossible to turn back, because sick people and parents of sick children have needs that cannot be denied. Of course applications of genetic manipulation must be carefully regulated so that they do not do harm. Regulations already exist and can be strengthened if necessary, but that does not mean that the useful applications of genetic manipulation can be stopped.

STS to FD:

7. To the statement "So I asked myself the age-old question, why does God permit war, why does God permit injustice…the problem of injustice seemed to me even more intractable than the problem of war" [*DU* p. 17] —how do

you define or describe "justice" (and therefore "injustice") in today's society?

FD to STS:

7. In today's society the most important problem is social injustice, which means the division of the population into rich and poor with unequal opportunities. We have legal justice but not social justice. In many ways the new technologies of the internet make social injustice worse, because people with access to the internet have access to jobs and business opportunities and information, while people without access are left behind and become unemployable. Genetic manipulation could also make social injustice worse, if genetic therapies and treatments are only available to the rich. This is the main danger that I see arising from genetic manipulation. The rules governing genetic therapies must be written so that they are equally accessible to rich and poor.

STS to FD:

8. Would you do anything different, knowing what you know now?

We thank you for sharing your insights with us through *Disturbing the Universe*...

Warm regards, STS Class

FD to STS:

8. I have done many stupid things in my life which I would avoid if I had the chance to live my life over again, but they are all personal things like spanking a child harder than necessary or forgetting my wife's birthday. So far as my public and professional life is concerned, I would not want to do anything differently. I think I used my talents as well as I could, as a scientist and as a writer. I think I made the right choice in ordering the priorities of my life, family first, friendships second, work third.

Happy Christmas and New Year to you all.

Yours sincerely,

Freeman Dyson

5 Not Jonah

But Jonah rose up to flee unto Tarshish...But the Lord sent out a great wind into the sea, and there was a mighty tempest...Then were the men exceedingly afraid, and said unto him... "What shall we do unto thee, that the sea may be calm unto us?" ...So they took up Jonah, and cast him into the sea; that the sea ceased from her raging. —*The Book of Jonah, Ch. 1*

The Spring 1999 semester was just beginning when a surprising letter landed in my mailbox:

> January 15, 1999
> Dear Professor Neuenschwander:
> We wish to consider Dr. Freeman Dyson for the Templeton Prize and your name has been given to us as a possible nominator....I would appreciate receiving your nomination of Dr. Dyson by March 15th.
> With my warmest good wishes,
> Yours sincerely
> Wilbert Forker,
> Executive Vice President, Templeton Foundation

The full name of the recognition is the Templeton Prize for Progress in Religion. The Templeton Foundation's purpose and programs are summarized by the Foundation's Vice President, Charles Harper: "What can we learn about life if we study the living world and its history from a perspective open to the dimension of purpose and meaning?"[1] I learned that recipients of the Templeton Prize for Progress in Religion—religion being broadly defined—have included Mother Teresa (1973), Cicily Saunders (1981), Alexander Solzhenitsyn (1983), Paul Davies (1995), Ian Barbour (1999), John Polkinghorne (2002), Jane Goodall (2021), Frank Wilczek (2022). The Foundation's board recognizes all faiths, e.g., the 1988 recipient was the Muslim activist Dr. Inamullah Kahn of Pakistan.

My first response on reading Mr. Forker's letter was "How can I possibly be up to meeting this request?" But in my mind's eye I saw the relief in all those students' faces when, thanks to Professor Dyson's input, they realized that science and religion offer two windows for looking at the world and both windows are needed. I recalled how Professor Dyson's chapter "The Argument from Design" [*DU* Ch. 23] helped us understand the distinction between the questions that belong to science and those that belong to religion, and how both are worthy of respect. So I had work to do. By March 15 the documents had been sent to Mr. Forker. They included the nomination form and a letter giving reasons for the nomination:

> In an overly extrapolated scientific reductionism one finds the view that science, and only science, holds the key to truth. The human being is seen as merely a complicated state of matter; there is no transcendent spiritual dimension to existence; and God must be dismissed as a fable of our own invention.
>
> At the opposite extreme, the acceptance of certain scientific paradigms becomes equated to atheism. The religious fundamentalist supposes that science, while useful for making the electric lights work, must at its root be an enemy of religious faith, especially in questions of origins or free will. The credibility of the larger religious community suffers with uninformed but vocal attacks against science.
>
> Somewhere between these extremes lies a large middle ground where a person may embrace religious and scientific perspectives simultaneously, honestly, and with intellectual integrity. Attempts to chart this middle ground will be taken seriously only if the cartographer commands the respect of the scientifically literate and the religiously sensitive. Professor Dyson is that cartographer.[2]

The nomination forms requested relevant quotations by Professor Dyson. These included some from *Disturbing the Universe:*

> *"It makes no sense to me to separate science from technology, technology from ethics, ethics from religion…"* [*DU* Ch. 1]

Other passages came from his speech to the conference of Catholic Bishops;[3] from passages in *Imagined Worlds*[4] and the *Reviews of Modern Physics* article "Time Without End."[5] The nomination letter's closing sentence observed,

> In reflecting over the sometimes turbulent relations between the scientific and religious traditions, of Professor Dyson it may be said, "Blessed are the peacemakers."[6]

This would not be the last we heard of Professor Dyson's consideration for the Templeton Prize.

The "Dyson Family Chronicle," the family's New Year's Letter for 1999, bore on our copy a hand-written greeting to the STS class: *"Happy New Year,… and thanks again for your friendly messages. Yours ever, Freeman."* Among other news, the Chronicle brought us up to date on daughter Mia's nursing and preaching careers. Mia's 1998 was impressive:

> *1998 was Mia's year. She graduated from the Princeton Theological Seminary in May, gave birth to a son in July, moved with her family back to their home in Maine in August, and was ordained as minister of Saint Andrew's Presbyterian Church in Kennebunk in November….*
>
> *The East-coasters reassembled in Maine for Halloween and All Saints' Day when Mia was ordained. On the day of the ordination, Mia was up early preparing her sermons, cooking breakfast pancakes for twelve, nursing Liam, and driving a car-load of kids and food to Kennebunk, before conducting the morning service. She preached two splendid sermons, a funny one for the kids and a serious one for the grown-ups. In the evening she was ordained, first taking her vows before the congregation in a ringing voice, then kneeling while the other ministers and elders laid their hands on her head. The congregation was friendly and welcoming. They had prepared a huge meal so that we could relax after the ceremony*

and get to know one another while the kids ran around. Mia's new life as a minister is off to a good start.[7]

In his Templeton speech the following year Professor Dyson was ecumenical: "When I am in Maine I am a Presbyterian, and when I am in England I am a Catholic." His sister Alice, living in Winchester, was a convert to Catholicism.

STS to FD:

6 April 1999

Dear Professor Dyson,

Once again our "Science, Technology, and Society" class has reduced about a hundred proposed questions to five...

It is a rare privilege for a class to communicate with its textbook author personally. Once again, on behalf of the nearly 50 students in our class, I thank you for contributing so much to our learning and to our lives.

Warm regards...

1. In "Aeropagitica" [*DU* Ch. 16] you describe how genetic research could be encouraged, yet society protected, provided that strict rules about applications are observed. We would appreciate your commentary concerning recent dramatic developments in bio-technology that could lead to especially wrenching ethical choices, such as "designer babies," organ cloning, etc.

FD to STS:

10 April 1999

Dear Dwight,

Thanks for your message, and thanks to the class for their good questions. It is a wet Saturday in Minnesota, with rain blowing horizontally all day long, a good day for staying indoors and answering letters. This morning I have been composing my homily for the chapel service on Monday morning. Whatever I have to say has to be said in seven minutes. That is good discipline. I send you the text of the homily as I wrote it this morning. As always, your comments and criticisms will be welcomed.

We interrupt Professor Dyson's letter to share the opening lines of his Gustavus Adolphus homily:

Two Views of the Future
Homily, given at Gustavus Adolphus College
Saint Peter, Minnesota, April 12, 1999

> *I don't know why you gave me the text of Jonah and the whale to preach from. I am a visitor here, enjoying the hospitality of the College, but I don't feel like Jonah. I am certainly not responsible for last year's tornado, and you people are not like the people on the ship going to Tarshish. Instead of finding a Jonah to blame for the tornado, you stood together and repaired the damage and got on with your lives. I am deeply impressed by the way you came through the disaster and took care of one another. You didn't need to find a Jonah to throw into the sea.*
>
> *That's all I have to say about Jonah....*

Professor Dyson's Gustavus Adolphus homily is reproduced in Appendix 1. Returning to his letter of April 10,

> *Now it's time to turn to your questions.*
>
> *1. What should we do to protect society from the harmful effects of recent developments in genetic engineering? The main thing I am doing here at Gustavus Adolphus is to run a seminar with twenty students studying precisely this question. We are using as our text the book: "Remaking Eden" by Lee Silver.*[8] *The students just finished reading the book. It is an excellent book, with detailed information about the fertility clinics where the technology of genetic engineering is applied to humans. The main facts are these: the fertility clinics are the most rapidly growing branch of medicine, they are not confined to rich countries but are growing rapidly in poor countries too, they are funded by private money and not by governments, and the driving force behind their growth is the desire of parents to have babies that they can call their own. The main danger to society is that, if the technology is available to rich parents and not to poor parents, it will widen the gap between rich*

and poor in a permanent and disastrous way, so that the children of the rich will become a hereditary caste with a monopoly of genetic advantages. The "designer babies" could become a hereditary upper class with the rest of the population condemned to inferior status.

At the Thursday meeting of the seminar we arranged a debate with the students arguing for and against three possible policies.

(1) Continue the present US policy, with a free market in genetic technology, unregulated except for the legal requirements of informed consent and FDA approval that apply to all medical treatments.

(2) Allow genetic manipulation of embryos only for the purpose of eliminating genetic deformities and life-threatening diseases, prohibiting manipulation for other purposes. So parents would be forbidden to use genetic manipulation to give their babies any advantages other than freedom from hereditary disease.

(3) Allow genetic manipulation of embryos for any purpose, but only when the resources exist to make it available to everybody. This means that genetic manipulation should become part of a public health service, like the polio vaccine that was distributed free to rich and poor alike as soon as it was available in quantity.

The division among the students was roughly, ten for policy (1), seven for (2), and three for (3). I am myself strongly in favor of (3), but this is clearly a loser in the eyes of the students. (3) only makes sense if the US has a national health insurance system, which most of our students do not consider a serious possibility. So long as the basic distribution of medical services is unfair, you cannot deal with genetic services in a fair way.

To summarize, the choice between the three policies is a choice of priorities. If you choose (1) you put personal freedom first. If you choose (3) you put fairness first. If you choose (2) you put tradition first. The students who supported (2) were mostly religious believers who consider genetic manipulation of embryos to be usurping the role of God. They would grudgingly allow it for the purpose of preventing disease, but no further.

I will be very much interested to hear how your students divide on this question. Although this is a Lutheran college, the majority of the students seem to be libertarian, believing the religion of the free market. For them the fertility clinic industry is just another example demonstrating the virtues of the free market. The freedom of parents to buy advantages for their children should apply to genes just as it applies to university education and day-care. The students who argued for (1) were the most articulate. The debate was mostly a battle between (1) and (3), with (1) prevailing. The students who supported (2) did not say much, and did not need to say much. Their beliefs are firm and do not need to be defended.

My discussion so far deals with designer babies. You also ask about organ cloning. We have talked about organ cloning but did not find that it raises any big ethical problems. In fact, if organ cloning becomes practical, it will alleviate the severe ethical problems that arise in the obtaining of organs for transplants under present conditions.

Several times the STS classes have conducted Professor Dyson's poll, using the same three options that he outlined in his Gustavus Adolphus seminar. Perhaps because our small southern university draws many of its students from a very conservative part of the country, over the years the results have overwhelmingly been in favor of option (2).

Voting Results for Professor Dyson's three categories of genetic policy:

Category	**1**	**2**	**3**
Professor Dyson's seminar, Spring 1999	10	7	3
STS classes:			
Spring 1999	7	35	0
Fall 2001	9	26	0
January 2004	7	31	1
January 2005	3	37	2
Spring 2005	1	17	5
Spring 2013	3	37	2
Spring 2014	1	33	0
Spring 2015	0	24	3

Comments:
Fall 2001: Upon further discussion, the students saw their support for 2 over 3 not as a choice between tradition and fairness, but as using the benefits of genetic technology while preserving diversity.

January 2004 mini-term: When the Gustavus Adolphus seminar votes were revealed the SNU students were surprised, but agreed with Professor Dyson's assessment that option 1 represents freedom; option 2, tradition; and option 3, fairness.

January 2005: After the voting one student suggested a fourth category, "No genetic manipulations at all." 13 students changed their vote to this option, all of them from the group that originally voted for No. 2.

These results are typical of every STS class that took the poll. Option 2 is the clear winner every time. This is a very traditional place, but the students who vote for No. 2 say their choices are not mostly about tradition. Professor Dyson noticed an irony in these responses. Let us briefly jump ahead in the chronology of the letters, to highlight some of his responses to our local voting on these three options:

FD to STS:

8 December 2001

Dear Dwight,

Thank you for the very interesting information about the student voting. It is interesting that they see genetic engineering as undesirable because it would promote uniformity of human populations, while I see it as dangerous because it would promote too much diversity. Perhaps they are right. But we will never know, unless we let the parents try it and see what they choose to do with it.

Happy Christmas and New Year to you all. Yours ever, Freeman

14 September 2013
Dear Dwight,

This is just to say thank you for the package of correspondence with your students that arrived yesterday…

I was very much interested in the result (3, 37, 2) of your vote [last spring] on genetic manipulation of babies…. There is a real difference between Oklahoma and Minnesota…

As always, I shall be glad to hear from the students. Yours ever, Freeman

Returning to 1999, that spring the United States and its allies intervened to stop "ethnic cleansing" in the war that was tearing Yugoslavia apart. The public was given the impression that the air power supporting NATO forces was especially potent. Few troops, it was said, would be needed on the ground. As students of "The Children's Crusade," [*DU* Ch. 3] we were skeptical.

STS to FD:

2. Would you care to comment on the events of the past two weeks between NATO and Yugoslavia, in view of your experiences at Bomber Command?

FD to STS:

2. Bombing of Yugoslavia. In my opinion this cannot possibly do any good. In the history of bombing there is only one clear case in which bombing succeeded. That was Japan in 1945, and it only succeeded because the bombing was totally destructive and ruthless. I hope the world today will not allow us to bomb Yugoslavia in that style. Even if we could do it, it would probably not succeed. In my opinion, Serbia has already won the war. This fact has nothing to do with right and wrong. No matter how right our objectives may be, bombing will not achieve them.

On the other hand, putting an army into Yugoslavia would be even worse than bombing. The worst danger is that the failure of the air-force will cause the army to believe that they could do it better.

Our third question refers to Professor Dyson's classification of civilizations, in the context of colonizing space, into three categories. [*DU* p. 212] A Type 1 civilization controls the biosphere of a planet; Type 2, a solar system; and Type 3, a galaxy.

STS to FD:

> 3. Given the fact that our lifetimes are so short compared to the time to traverse interstellar distances, do you think we will ever develop a Type 3 civilization? What would motivate people to begin such a migration knowing that only their descendants so many generations removed would finish it?

FD to STS:

> 3. *Type 3 civilizations. These necessarily take hundreds of thousands of years to grow, since it takes almost a hundred thousand years for light to travel from one side of a galaxy to the other. When you are talking about things that take hundreds of thousands of years to do, the present human lifetime is not relevant. Long before we are in a position to evolve a Type 3 civilization, we will probably know how to control our life processes so that we can live as long as we like. Many of us might prefer not to be immortal, but there will be a variety of kinds of people with a variety of lifetimes. People who want to travel long distances will be able to choose appropriate lifetimes. Of course it is impossible to guess what their motivations might be. Why did the paleo-native Americans travel all the way from Alaska to Tierra del Fuego in less than a thousand years?*

STS to FD:

> 4. We have been impressed with your many interesting life experiences (e.g., being present at Dr. Martin Luther King's "I Have a Dream" speech) [*DU* pp. 140–141] in addition to your scientific accomplishments. On what projects are you currently working, and what have you not yet done that you would still like to do?

FD to STS:

4. Projects. I just finished the revised and improved second edition of "Origins of Life" and sent it off to the printer. That has been my main project for the past year. It will be published by Cambridge University Press sometime this year.[9] *Now, when I am finished with this visit to Gustavus Adolphus, I will go back to Princeton and think about what to do next.*

I am now 75 years old and have been retired for five years. These five years have been mostly spent traveling and teaching, visiting many places in the same way as I am now visiting Gustavus Adolphus. I have enjoyed these visits but I have spent too little time at home. I hope to stay home more in the next five years, see more of the grandchildren and perhaps write another book. That is all I can say about future plans.

STS to FD:

5. What would you say is the greatest invention of mankind?
Thank you, The STS Class

FD to STS:

5. What is the greatest invention of mankind? Language. After that, domesticating plants and animals. After that, writing. After that, religion. After that, science. I am in the middle of reading an excellent book, "Guns, Germs and Steel: the Fates of Human Societies," by Jared Diamond.[10] *I recommend it to you and your students. It deals with these questions on a very broad basis: which inventions were most important and why some societies made them and others did not. Diamond is a field biologist who spent a large part of his life among the primitive people in New Guinea and elsewhere. He says he knows by direct experience that these people are at least as smart as we are. The fact that we made the important inventions and they did not has nothing to do with our being smarter. We were just luckier. It was accidents of geography that gave us our advantage. I think he proves his case very convincingly.*

That's enough for today. Thank you for reducing the questions from a hundred to five. All good wishes to the class. And to you, from Freeman

6 The World Soul

My mother did not like the phrase Cosmic Unity. It was too pretentious. She preferred to call it a world soul. She imagined that she was herself a piece of the world soul that had been given freedom to grow and develop independently so long as she was alive. After death, she expected to merge back into the world soul, losing her personal identity but preserving her memories and her intelligence. Whatever knowledge and wisdom she had acquired during her life would add to the world soul's store of knowledge and wisdom. —*DU, p. 252*

For the Spring 1999 semester our campus bookstore was unable to find copies of *Disturbing the Universe,* new or used. Ever resourceful, the problem was temporarily solved with a legally photocopied special edition of *DU.* We sent Professor Dyson a copy signed by all the students, along with a note:

STS to FD:

May 3, 1999

Dear Professor Dyson,

We held a book signing party in our class today, in your honor. Thank you for being part of our class. Our university bookstore manager obtained permission from Harper Collins to reproduce *Disturbing the Universe.* Maybe this will encourage the publisher to re-print it!

Best wishes to you and your family, the STS class

Our next letter exchange occurred as Christmas 1999 approached.

STS to FD:

7 December 1999

Dear Professor Dyson,

On behalf of the Fall '99 STS class, we wish you and your family a joyous and meaningful Christmas Season. We hope you will all have a wonderful time together....

As is our custom (one we hope has not grown burdensome to you), the students have proposed a few questions....

The first question emerged from our discussion of the Strategic Defense Initiative that was announced by President Ronald Reagan in 1983. We read an article that had been published in the June 1985 issue of *Physics Today* by Wolfgang Panofsky. He argued that the SDI program "is too large, too political, raises false hopes and poses grave dangers to national and world security."[1] In view of Professor Dyson's articulation of the ethics of defense, the class raised an SDI question with him:

STS to FD:

1. We don't want to sound like the crazy general in "Dr. Strangelove," but is there a possibility of having an affordable Strategic Defense Initiative? Was SDI consistent in spirit with the stand you make in "The Ethics of Defense?" [*DU* Ch. 13]

FD to STS:

8 December 1999

Dear Dwight,

The questions from your students go immediately to the top of the pile of unanswered letters. The others can wait for a few days longer. Please give the students my greetings and thanks. Here are some off-the-cuff answers...

1. Yes, I agree there is a possibility of an affordable missile defense, and this is consistent with the defensive strategy that I was advocating in the book. But this only makes sense, politically and morally, if it is combined with drastic reductions in offensive forces. To build a missile defense now, when we have no intention of reducing offensive forces seriously, would be wrong from every point of view. It would be seen by the Russians as threatening, it would do nothing to counter the real danger of terrorist bombs arriving by boat or by airplane, and it would be a great waste of money. I would say, let's get rid of at least ninety percent of our offensive missiles first, and then missile defense might be part of a genuine shift to a defensive strategy.[2]

In our class discussions on environmental sustainability we see how exponential growth in the human population drives increasing energy consumption, the diminishment of natural landscapes, habitat loss and species extinction, increasing pollution, climate change, and so on. It's not just *what* we do, but the *rate* at which we do it that matters.

STS to FD:

2. Do you think the world will naturally take care of overpopulation through plagues and starvation? Medicine continues to prolong life—what do you think the expected life span for humans will be by the year 2050?

FD to STS:

2. I was delighted to read recently that the average number of children in Mexican families dropped from 7 to 2.5 in the last thirty years. In Italy the birthrate is so low that the Italians are saying, in a hundred years there will be nobody here but Albanians. Somehow the women of the world are getting the message, you don't have to have a lot of babies. This is the cure for over-population, not plagues and starvation.

Medicine increases the average life-span but does not do much to increase the upper limit. Sophocles wrote his last play at the age of ninety. Ninety remains the upper limit for practical purposes. I hope this remains true. The worst thing I can imagine would be if the doctors find a cure for death. On Sunday we went with our grandsons to sing carols to the patients in the local nursing home, most of them ninety-year-olds stuck there until they die. They loved the boys and the carols, but it is a sad and depressing place. Living longer than ninety is not much fun.

When he wrote this, Professor Dyson was a week from his 76th birthday. As it turned out, like Sophocles he was still writing in his nineties, still going to the office.

STS to FD:

3. Why would cloning be considered when we're already worried about over-population? What are the long-term implications of cloning for the diversity of humans?

FD to STS:

3. Cloning of humans is not a big deal. It may be a preferred option for a small fraction of parents who are unable to get babies by other methods, but it will not change the total population significantly.

The serious problems raised by genetic engineering do not come from cloning but from the genetic manipulation of human embryos. Many parents, not only those who have fertility problems, may want to give their babies a better chance in life by putting in a few extra good genes or taking out a few bad ones, as soon as the technology is available. This could be a severe problem, especially if the good genes are only available to rich parents. The children of the rich could quickly become a hereditary caste. I recommend the book, "Remaking Eden" by Lee Silver,[3] *who discusses the possibilities and the dangers realistically. As Lee Silver remarks, fertility clinics are a huge and profitable branch of medicine all over the world, and the technology is driven forward and paid for by the parents, not by the governments.*

STS to FD:

4. You have met many interesting people in your life. You have described what you learned from them. Flipping the coin, what is one thing you think people have learned from you that influenced them significantly?

FD to STS:

4. I can't answer this one. When I talk or write, I am the sower throwing seed into the wind. I never know where it lands or where it may be sprouting.

Professor Dyson's response reminds us that wisdom leaves it to others to describe one's accomplishments and influence. Professor Dyson may see himself as a sower throwing seed into the wind, but we have seen his influence flower among people he never met, and not only STS students. Another example includes Marcelo Gleiser's book *The Dancing Universe: From Creation Myths to the Big Bang.* Gleiser writes in his preface, "Freeman Dyson's initial encouragement was really crucial to set me in the right direction."[4] And in Christa Tibbit's book *Einstein's God,*[5] which consists of annotated transcripts of conversations she conducted on her radio program *On Being,*[6] Tippett's first conversation in the book features Professor Dyson. Closer to home, as mentioned earlier, our sociology colleague, Professor Ron Wright, obtained a Templeton Foundation grant to hold a series of "town meetings" on science and religion. The events were called "Two Windows" seminars[7] where Professor Dyson's inspiration for the term was publicly acknowledged. In the early 2000s the STS class also studied a *The Unholy War: The Conflict Between Science and Religion* by Karl Giberson.[8] The Fall 2000 semester STS class sent Professor Giberson a few questions. He prefaced his answers with the comment "I am humbled to respond, knowing that Freeman Dyson is also a part of your course…."[9]

The next question was motivated by Chapter 23 of *DU,* "The Argument from Design." There Professor Dyson describes "two separate levels" on which "mind enters into our awareness of nature." At the first level, we are aware of our own minds. At the next level, we are aware that "the mind of an observer is again involved in the description of events." [*DU* p. 249] An example of this second level of awareness occurs when one performs an experiment with an electron. Whether it behaves like a wave or like a particle depends on how one interacts with it, suggesting the existence of some sort of linkage between the electron and our minds. Professor Dyson identifies a third level, namely, conditions in the universe that make the existence of minds possible. He speculates on a connection between the three levels. Recalling how the conditions necessary for life as we know it require narrow ranges in the values in the fundamental constants, he writes

> *The architecture of the universe is consistent with the hypothesis that mind plays an essential role in its functioning…. The peculiar harmony between the structure of the universe and the needs of life and intelligence is a third manifestation of the importance of mind in the scheme of things. This is as far as we can go as scientists. We*

> *have evidence that mind is important on three levels. We have no evidence of any deeper unifying hypothesis that would tie these three levels together. As individuals, some of us may be willing to go further. Some of us may be willing to entertain the hypothesis that there exists a universal mind or world soul which underlies the manifestations of mind that we observe....The existence of a world soul is a question that belongs to religion and not to science.* [*DU* pp. 251–252]

Most religious traditions are centered on the hypothesis of God or a Cosmic Mind, Great Spirit or a World Soul. A fundamental article of faith of the Christian religion assumes the bodily resurrection of Christ after his crucifixion. If that was an historical event, it means the existence of a Divine Mind that interacts with the physical world but operates outside of nature, and therefore lies beyond the scope of science. Hence the following question:

STS to FD*:*

> 5. Your description of "mind" was very thought-provoking. In view of the metaphysical possibility of an "ultimate mind," do you believe in the possibility of the supernatural, e.g., the resurrection of Christ?

FD to STS*:*

> *5. To me, religion is not a matter of belief but a way of life. I go to church to be part of a community of caring people. I consider myself a Christian, but I don't believe in the resurrection. I am not saying dogmatically that the resurrection is impossible. Obviously there are more things in heaven and earth than we are capable of understanding, and this may be one of them. Being a scientist means being comfortable with uncertainty.*

The question, "Which is more important: what you believe, or how you behave?" arises frequently during our class discussions about religion. The stability of human societies requires a deep appreciation of uncertainty. As columnist George Will observed, "The greatest threat to civilization is an excess of certitude."[10] Another author we often read,

Jacob Bronowski, made this point passionately as he stood by the Auschwitz pond into which were flushed the ashes of four million people, victims of the Holocaust: "When people believe that they have absolute knowledge, with no test in reality, this is how they behave.... Science is a very human form of knowledge....Every judgment in science stands on the edge of error, and is personal. Science is a tribute to what we can know *although* we are fallible."[11]

Most chapter titles in *Disturbing the Universe* borrow a theme from literature or music, such as "The Redemption of Faust" (Ch. 2), "Prelude in E-Flat Minor" (Ch. 8), or "The Island of Doctor Moreau" (Ch. 15). References to literature and poetry are found throughout in the book. Professor Dyson's father, Sir George Dyson, was a distinguished conductor and composer (his best-known work is *The Canterbury Pilgrims*); his mother, Mildred, trained as a lawyer, memorized many poems and was well-read in literature.

Fig. 6.1. Freeman Dyson's parents: Sir George (1883–1964) and Lady Mildred Lucy Atkey Dyson at a young age (1897–1974). Photos courtesy of Imme Dyson.

STS to FD:

6. Along with the science and technology, your book displays an intense passion for the humanities... What led you to

develop such broad interests? Do you view the humanities and sciences as separate disciplines, or as two complementary sides of the same coin?

Many other questions were asked that were similar to those asked by previous classes.... For example... there were at least three questions submitted about the meaning of the book's final scene with the throne of God and the baby—that scene, and its implications, continues to fascinate us, and you responded to us about it earlier.

We hope that you and all your loved ones are well. Best wishes and Season's Greetings.

Warm regards, the STS Class

FD to STS:

6. I emphatically agree with the statement that the humanities and sciences are complementary sides of the same coin. They are different ways of looking at the same world. I was always interested in literature and poetry, not so much in music and philosophy. My favorite teacher in high-school was a chemistry teacher who read poetry aloud to his class. As he said, we didn't need to come to class to learn chemistry.[12] *And my parents were both humanists.*

That's all for today. I end with some good advice which we got from our seven-year-old granddaughter Bryn in Maine. She wrote a letter for my wife's birthday. Here it is.

"Dear Omi, I Bryn hope and wish for you a very happy birthday! I hope that you have a good time at your birthday celebration. Remember that there are more years to come and I hope you spend them wisely! I love you, Bryn."

Happy New Millennium to all of you,

Yours, Freeman Dyson

7 Not With Words, But With a Smile

Sweet dreams form a shade
O'er my lovely infant's head...
Sweet smiles in the night,
Hover over my delight...
Sweet babe in thy face,
Holy image I can trace...
—William Blake, *"A Cradle Song" excerpts*

February 1, 2000
Dear Professor Neuenschwander:

It is a pleasure to write to you with the good news that the recipient of the Templeton Prize for the year 2000 is Dr. Freeman Dyson...

A Media Conference is planned for 11 a.m. on Wednesday, March 22 at the Church Centre of the United Nations, 777 UN Plaza in New York...

The public ceremony at which Dr. Dyson will deliver his acceptance speech will be held at 7 p.m. on Tuesday, 16th of May at the Washington National Cathedral... It would be appreciated if you would keep the name of the recipient in confidence until March 22.

Yours sincerely,
Wilbert Forker, Executive Vice President

Despite the generous comment that Professor Dyson made in his letter to me of March 29, 2000 (below), I do not claim that my letter of nomination made any difference in securing the Templeton Prize for Professor Dyson. I suspect that the Templeton board already had Professor Dyson in their sights to honor in this way, and the rules of the prize required somebody from outside the Foundation to write the necessary nomination letter. I'm sure that many others could have been asked to write a letter, and perhaps they did. Be that as it may, it was a privilege to have a small role in the process, and I was glad to do it. By the spring of 2000 our STS classes had corresponded with Professor Dyson for seven years, and the Templeton nomination gave us an opportunity to publicly express our gratitude to him.

FD to DN:

29 March 2000

Dear Dwight, what can I say? This absurd and undeserved glory that has overwhelmed me in the last few days is entirely due to you. I am of course immensely grateful. I deeply respect Sir John and the purpose he has in mind for these prizes, and so I accept his largesse with gratitude. Gratitude to him and to you. And still I think, this is a bit crazy.

You may have found it odd that I was totally surprised by the award, when after all I knew that you had nominated me and you sent me a copy of your nomination letter. What happened was this. After you sent me your nomination letter, Ian Barbour won the prize. So I thought, that's fine, Ian is a much better candidate than I am, and so naturally he won the prize, and that's the end of it. I never imagined that your nomination might be held over until the following year. I thought it was dead. So I was totally surprised when they told me I had won the prize a year later.

Now of course everybody asks me, what are you going to do with the money? I say, I don't know, and that is true.... We have daughters who are working mothers with small children, and they need all the help we can give them. For us, family comes first. Above all, we don't want to be a burden to our children when we are old and decrepit. We shall probably give a modest slice of the money to charity...the problem is to identify the charities where a modest amount of money can do the most good. Do you have a particular good cause that you would like us to support? If so, please let us know.... Of course we don't promise to follow your advice, but we shall at least listen to it carefully.

All good wishes and thanks to your students as well as yourself. I hope we shall meet you in Washington in May.

Yours ever, Freeman

P.S. I forgot to say, what were you doing in Costa Rica? That sounds interesting, and I hope to hear about it whenever you have time. Yours again, Freeman

DN to FD:

29 March 2000:

Dear Professor Dyson,

The short answer is this: I went to Costa Rica, at the request of my biology colleague Leo Finkenbinder, to teach astronomy... I must explain, which brings up the long answer...

To cut the long answer short, through a history of personal relationships between SNU biology professor Leo Finkenbinder and his wife Zana, and the family of Efrain Chacón, our university established a field station, called the Quetzal Education Research Center, located in the Rio Savegre valley of the Talamanca Mountains in Costa Rica. Students, researchers, and bird watchers come to the QERC from around the globe to hike the narrow cloud forest trails to study the fabulous biological diversity, including the Resplendent Quetzal, who thrives here in one of its few remaining habitats. Our students take courses in this spectacular setting where, according to the biologists, in a hundred yards of hiking one passes by more biodiversity than exists on the entire North American Continent. Here the field biologist Leo Finkenbinder and I team-taught a course that we named "The Astronomical Basis of Life on Earth." One can describe the course as astrobiology looking inward, examining astronomical conditions that are necessary for life as we know it to exist.[1] For example, after discussing the nuclear reactions that power the Sun, and considering the meaning of the habitable zone about a star, we hike through the forest where Leo points out the diverse light-gathering strategies of photo-synthetic organisms. To allow light to reach the lower leaves, some plants arrange their leaves in a spiral; the leaves of other species have large holes, and so on. This setting offers a beautiful segue into a larger awareness of environmental sustainability. This oasis planet does not need us, but oh how we need it! Even when the human race expands across the solar system and beyond, this planet will still be occupied by our descendants.

In our spring break trips to our university's field station in Costa Rica, it was obvious that the greatest obstacle to participation for many students was travel expenses, an additional cost beyond the usual tuition and fees. After some discussion on our end, we suggested to the Dyson family the establishment of the "Dyson Travel Scholarship" explicitly for students who wish to study at the QERC. The arrangements were set up through our University Advancement office. Managed by the biology department, students taking a course or conducting a research project that includes travel to the QERC are welcome to apply for the Dyson Scholarship. Professor Dyson's generous offer to share some of the Templeton Prize funds has been a tremendous blessing to our students and their families, encouraging a growing awareness of environmental sustainability issues, while offering to the students networking opportunities and indelible life experiences.

The award for 2000 of the Templeton Prize

for Progress in Religion

Washington National Cathedral, Tuesday, 16th May, 2000

Organ Prelude — Prelude and Fugue in G Major, J S Bach

Cathedral Choirs of Girls, Boys and Men — Let us now praise famous men, Leo Sowerby; Ye that do your Master's will, George Dyson

Opening Prayer...The Reverend Dr. Lloyd Ogilvie
Chaplain to The United States Senate

Welcome...The Very Reverend Nathan D. Baxter
Dean of Washington National Cathedral

Address...The Honorable James H. Billington
Librarian of Congress

Address...Sir John Templeton
Founder of the Templeton Prize

Cathedral Choirs — I will worship toward Thy Holy Temple, George Dyson

Address...Dr. Freeman J. Dyson
The Recipient for 2000

Cathedral Choirs — Confortare, George Dyson

Closing Prayer...The Right Reverend Ronald H. Haines
Bishop of the Diocese of Washington

Organ Postlude — Toccata, Leo Sowerby

Fig. 7.1. Program for the 2000 Templeton Prize acceptance speech ceremony, 16 May, 2000, held in National Cathedral, Washington, DC.

When Professor Dyson was awarded the Templeton Prize for Progress in Religion, HRH the Duke of Edinburgh made the formal presentation on May 9, 2000, in a private ceremony held at Buckingham Palace. Professor Dyson's acceptance speech was delivered one week later at Washington National Cathedral (Fig. 7.1).

Before the acceptance speech ceremony, a reception was held for Professor Dyson and his family in the St. Albans School on the Cathedral campus. Professor Dyson and Imme stood in a receiving line as many distinguished people patiently waited to speak with them. I was standing off the side watching these proceedings. Suddenly a swarm of young children (Fig. 7.2) burst into the room, running straight for Professor Dyson with joyful shouts of "Papa! Papa!" Professor Dyson immediately turned towards these boisterous children and knelt down, welcoming them with open arms as they swarmed over him. The guests in the receiving line had to wait. But no one seemed to mind. We all enjoyed witnessing this special moment between a beloved grandfather and his delightful mob of merry grandchildren.

Fig. 7.2. The Dyson East Coast grandchildren who attended Professor Dyson's Templeton Award acceptance speech. L to R: George, Donald, Tess, Bryn, Liam, Randall. At the time of the speech the youngest, Liam, was a toddler. Photo courtesy of the Dyson family.

Later, as everyone migrated into the Cathedral for the ceremony, I happened to be standing in the vestibule by a couple who introduced themselves as neighbors of Freeman and Imme Dyson. They invited me to sit with them. Our conversation before the ceremony led later to another letter to Professor Dyson. The ceremony's prelude music was a Prelude and Fugue by Johann Sebastian Bach, but the music performed during the ceremony itself were compositions by Sir George Dyson, Freeman's father. As the ceremony opened, the Cathedral rang with George Dyson's music as robed clerics marched in high procession down the center aisle of the nave. Professor Dyson walked casually behind them, smiling and discreetly waving to people he knew, as if he had accidently wandered into the procession. It was delightful to witness. In the midst of all this pomp and circumstance, Professor Dyson enjoyed merely being himself.

But when he stood up to deliver his acceptance speech, his voice rang out to resonate the space within the entire Cathedral. He began by acknowledging that former Templeton Prize recipients were either "saints or theologians." But, he said, he was neither a saint nor a theologian. His speech is reproduced in Appendix 2. Here is an excerpt:

> *Science and religion are two windows that people look through, trying to understand the big universe outside, trying to understand why we are here. The two windows give different views, but they look out at the same universe. Both views are one-sided, neither is complete. Both leave out essential features of the real world. And both are worthy of respect.*
>
> *Trouble arises when either science or religion claims universal jurisdiction, when either religious dogma or scientific dogma claims to be infallible. Religious creationists and scientific materialists are equally dogmatic and insensitive. By their arrogance they bring both science and religion into disrepute....*
>
> *The great question for our time is, how to make sure that the continuing scientific revolution brings benefits to everybody rather than widening the gap between rich and poor. To lift up poor countries, and poor people in rich countries, from poverty, to give them a chance of a decent life, technology is not enough. Technology must be guided and driven by ethics if it is to do more than provide new toys for the rich... Science and religion should work together*

to abolish the gross inequalities that prevail in the modern world. That is my vision...[2]

Before the Templeton event in Washington D.C., the Spring 2000 STS students prepared more questions for Professor Dyson. After introductory salutations the questions began:

STS to FD:

1 May 2000

Dear Professor Dyson,...

1. *In Disturbing the Universe* you said, "Robert Oppenheimer was driven to build atomic bombs by the fear that if he did not seize this power, Hitler would seize it first. Edward Teller was driven to build hydrogen bombs by the fear that Stalin would use this power to rule the world." [*DU* p. 91] Do you feel that a person's drive to seize power always comes from fear?

FD to STS:

1 May 2000

Dear Dwight,

Luckily I have a few short days left before your term ends. The relentless Templeton time-table begins again with a new series of interviews in London and Washington....

1. No, I don't think that fear is usually the main reason for seizing power. You might say that Oppenheimer and Teller were both unsuccessful in seizing power because they were driven by fear more than by a natural love of power. The power they seized did not last. The people who seize power successfully are usually charismatic leaders who love power and are not afraid of anybody. Examples we have seen in the last century were Lenin, Hitler, Castro. They believed that they were fulfilling a higher destiny and therefore did not need to be afraid. Not only the evil leaders had this unafraid quality. Good leaders like Joan of Arc, George Washington, Gandhi, had it too. They also believed in a higher destiny and were not afraid when they seized power.

STS to FD:

2. In Ch. 20 ("Clades and Clones") you discussed languages, and how some are dying off. What do you feel about we Americans? – here we are, an influential country, yet we do not put emphasis on learning other languages besides English.[3]

FD to STS:

2. *Yes, it is a great shame that Americans generally do not take the trouble to learn other peoples' languages. But I have to admit that the English are even worse. I grew up in England and never learned to speak any other language decently.*[4] *I admire the Mormons, because they spend two years abroad as missionaries and take the trouble to learn the language of the country they are sent to. When I visited Brigham Young University in Utah, I was delighted to find a crowd of students who all knew a second language and were familiar with other cultures. But you don't have to be a Mormon to spend two years in a different culture. You can also join the Peace Corps. My Peace Corps daughter perfected her French in francophone Cameroon.*

Of course the best time to learn a second language is when you are two or three. But you lose it if you do not keep on speaking it as you grow up. I wish the schools in the US would allow (not compel) children to be immersed in a second language as soon and as long as possible. This would be easy in regions where Spanish speakers are a big fraction of the population, as they are in New York or California. It is a terrible waste to bring up children surrounded by Spanish-speakers and not speaking Spanish.

STS to FD:

3. What causes children to lose their fascination with science? One student writes,

"I remember loving science in my first few years of school…I looked forward to the field trips to science and marine centers. My sisters and I would beg our parents to take us there when we were not in school. However, now that I am older, I can hardly stand to sit through a science class. I even changed my major so I would not have to take so many science hours. From talking to students, it sounds like this is a common feeling among many. Looking back, I cannot remember a specific time or class that made me lose my interest in science. I know that not everyone is going to be a scientist, but there should be at least some tolerance for it."

Do you think it is just a natural part of growing older, that one forgets how to be curious about the world? Or does the way science is presented in schools cause children to turn away from it? Do you have any suggestions for keeping children interested in science?

FD to STS:

3. Yes, the way you describe it rings true. Children are naturally curious up to the age of ten and then lose it. A month ago I was teaching a class of fourth-graders and they were wonderful, full of curiosity and enthusiasm. I know that if they had been seventh-graders a lot of them would have been bored. Every science teacher knows this. I don't blame this on the way science is taught in schools, because it happens even when science is taught well. Certainly kids are turned off even faster when science is taught badly, but many of them get turned off anyway.

I don't think science is different from music or drawing in the way children grow out of it. My grandsons are in an elementary school with an excellent art teacher. We recently went to the exhibition where the kids show their stuff. Amazingly good paintings and drawings done by hundreds of kids. The art teacher knows every one of the 500 kids in the school by name. A large majority of them do great stuff up to the age of ten. But then, as they become teen-agers, most of them lose it. Only a small minority keep on drawing and painting and keep their talent. And it is the same with science. And with music and mathematics. I think it is part of human nature that we start out as children interested in everything, then become specialists as young adults, and then broaden our interests again in later life. There is nothing wrong with being bored by science if you don't happen to be scientifically gifted. What is wrong is for teachers to have to try to stuff science into people who are bored with it.

So my recommendation for education in science is, give it to young children who take to it naturally, and to old people who come to it fresh in later life, and make it optional for teen-agers and college students, just as we do with art and music. The idea of forcing everyone to be "scientifically literate" does not work. As Tolstoy said a hundred and fifty years ago, "All that the greater part of the people carry away from school is the horror of schooling."[5] *So it is no surprise that the attempt to force people to think scientifically gives them a horror of science. Most of us are just not built that way.*

"Cosmic Unity" was Freeman's youthful philosophy of life. It held that "There is only one of us. We are all the same person." [*DU* pp. 17–18]

STS to FD:

4. Another student writes, "I had a hard time wrapping my mind around the concept of Cosmic Unity and how it was significant in regard to ethics. Could you explain a little more how it was 'a firm foundation for ethics' and how you came to realize this concept?"

FD to STS:

4. My idea about Cosmic Unity was very simple. You and I and Hitler and Stalin are the same person in different disguises. When you hurt another person you are hurting yourself. It makes a firm foundation for ethics. Since each person you meet is you, it makes sense for you to treat them as you would like to be treated yourself. It makes no sense for you to treat them badly. In other words, the golden rule follows: "Love thy neighbor as thyself," or "Do as you would be done by."

I thought of Cosmic Unity first as a way to solve the problem of injustice in the world. If we are all the same person, then there is no injustice, because we are all suffering equally. So injustice does not really exist and is not a problem. Then, after having invented Cosmic Unity to solve the problem of injustice, I realized that it also solves the problem of ethics. If you kill your enemy, you are killing yourself, so it follows that you should not kill.

Of course this all sounds much too simple to be true. When I was fourteen I believed it whole-heartedly. It helped me to deal with the tragic world of the nineteen-thirties in which I grew up. Now I think of Cosmic Unity as a hope rather than a fact. It still makes sense as a foundation for ethics. It is an ideal toward which we should strive.

STS to FD:

5. In your opinion, did President Truman make the right decision to drop the atomic bombs on Hiroshima and Nagasaki?

FD to STS:

5. Yes, I think Truman made the right decision. So long as World War 2 continued, we were killing more people every month than we killed at Hiroshima and Nagasaki. So the dropping of the bombs was saving lives, if it shortened the war only by a couple of months. We can never know whether or not Japan would have surrendered in a couple of months if the bombs had not been dropped. The best evidence on this question was collected by Robert Butow in his book "Japan's Decision to Surrender."[6] *Butow interviewed the Japanese wartime leaders while their memories were still fresh and asked them the question "Would Japan have surrendered without the dropping of the bombs?" The Japanese leaders themselves did not know the answer. Therefore we cannot know the answer, and certainly Truman could not know the answer. It is quite possible that if the bombs had not been dropped the Japanese armies would have fought to the last man as they did in Okinawa in the spring of 1945, with millions more dead before the war ended. Nobody can know whether or not this would have happened. So I say, Truman made the right decision.*

Another question is, whether Truman in fact could have made a decision not to drop the bombs and made the decision stick. Perhaps he had no choice. A decision not to drop the bombs would have been violently opposed by the army generals and their friends in Congress, and Truman was in a weak position to impose his will on Congress, having only recently become President. Certainly Truman could only have prevailed over Congress if he had overwhelmingly strong reasons not to drop the bombs. And I am sure he did not see any strong reason not to take this chance to end the war. I thought at the time that he was right, and I still think so.

STS to FD, a postscript to the numbered student questions:

P.S. Eight of the students again asked a question about the dream with which you close *Disturbing the Universe*. [*DU*, Ch. 24] Those questions took the form of asking about the symbolism of the baby, and what questions you wanted to ask when you made the appointment. I shared with the class your responses to those questions from previous classes. We thought you might be interested to see how this image continues to capture the imagination of readers.

FD to STS:

P.S. About the baby, I don't need to add much to what I said before. The symbolism is obvious. My father who was a musician used to say that music was the closest he ever came to God. And I feel the same way about babies. Recently I was at a party in California and had the luck to hold a baby for two hours while his mother cooked the dinner. His mother was grateful and so was I. Holding a baby is the closest I ever come to God. So the dream made sense. I think of God as coming into existence like a baby, still close to the beginning, no end in sight.

You ask what questions I had in mind to ask him. Just the usual questions, why the world is so full of misery and injustice and evil, why so many people never have a chance of a decent life, why the people who have the power to make things better are mostly viscous and stupid. If God made us, couldn't he have made us better? And of course the questions were answered the same way babies always answer questions, not with words but with a smile.

I guess that's all for this time. Please thank the students again and tell them I'll always be glad to hear from them individually.

Good wishes for summer jobs and travel to all of you. Yours ever, Freeman

After the summer, the Fall 2000 class came through with a few questions:

STS to FD:

4 December 2000

Dear Professor Dyson,

Our Science, Society, and Technology class at SNU wishes you and your family a joyous Christmas season... In the music that plays as the students gather before class we have sampled "The Canterbury Pilgrims" by George Dyson. We'll see if it soothes the mind as students gather soon for the final exam!

1. If you were beginning your scientific career today, and were in the process of choosing among the various sub-fields that are hot on the burners right now, to which one(s) would you choose to contribute your time and talents?

FD to STS:

7 December 2000

Dear Dwight, thanks for your message. Delighted to hear you are playing bits of the Canterbury Pilgrims. Please tell the students, thanks for their questions and you will hear from me soon with the answers. This time the students' list of questions is shorter than last year. I suppose you told them to avoid repeating last year's list, and now they are beginning to run out of questions. Anyhow, here are my answers. Since there are fewer question, my answers will be a bit longer.

1. The problem I had with choosing a field of science to work in was that my interests were much broader than my talents. I have always been intensely interested in biology and medicine, but I couldn't do biology or medicine because I didn't have the right skills. In choosing a profession, the first criterion has to be, do you have what it takes? I didn't have what it took to be a biologist. My only real talent was mathematics, so I became a mathematician. Then I found that with my mathematics I could also solve problems in physics, so I switched to physics. I became a theoretical physicist. In biology there was much less scope for mathematics.

Things are changing, and now it is possible to be a theoretical biologist, but real biologists who work in the laboratory do not have much respect for theoretical biologists. If I were young now, I might have become a theoretical biologist, studying the evolution of genomes and populations with mathematical tools. But I think it more likely that I would have stuck to mathematics and physics. Probably I would have tried my hand at string theory, the fashionable branch of theoretical physics that uses very sophisticated mathematics. String theory may not have much to do with the real world, but it is mathematically deep and beautiful. I am sure I could have enjoyed a happy life as a string-theorist. Whether or not string theory is useful, it provides the right sort of sand-box for a person with my talent to play in. And after making a reputation as a string-theorist, I could then have broadened my activities and written books about other things as I actually did.

STS to FD:

2. What is our most important duty as individual human beings?

FD to STS:

2. I don't believe there is any duty that is most important for everybody. People have different circumstances, different opportunities and different talents. Therefore they have different duties. A duty is like a vocation. My daughter Mia had a vocation to be a Presbyterian minister. Now she has a church and it is her duty to help her parishioners by preaching and counseling and administering the church. That is for her the most important duty. But I don't have a duty to be a minister. My vocation, if I have one, is different. At present, after I have retired as a working scientist, my most important duty is to help my wife take care of children and grandchildren.

For you young students, there is a wide choice of vocations and a wide choice of duties. Some of you have a duty to be Martha and some have a duty to be Mary. I always thought Jesus was unfair when he scolded Martha for being jealous of Mary. The world needs Martha just as much as it needs Mary.

We recall here the biblical story about Mary and Martha as told in the Gospel according to Luke:

> As Jesus and his disciples were on their way, he came to a village where a woman named Martha opened her home to him. She had a sister called Mary, who sat at the Lord's feet listening to what he said. But Martha was distracted by all the preparations that had to be made. She came to him and asked, "Lord, don't you care that my sister has left me to do the work by myself? Tell her to help me!"
>
> "Martha, Martha," the Lord answered, "you are worried and upset about many things, but few things are needed—or indeed only one. Mary has chosen what is better, and it will not be taken away from her."[7]

Returning to Professor Dyson's letter,

Mia knows this well, since she is also a mother with three children. As a mother she is Martha and as a minister she is Mary. She knows that, as Jesus said, her first duty is to be Mary. Her second duty is to be Martha and not to be jealous of Mary. The first sermon that she preached when she was ordained was on the story of Martha and Mary. She told us how she has to be Mary first and Martha second. At some point in your lives, you will all be faced with similar choices between conflicting duties. There is no easy way to decide which duty comes first.

STS to FD:

3. What other works was your mother fond of quoting besides Chaucer's works, Goethe's *Faust* and Terentius Afer's *The Self-Tormenter*?

FD to STS:

3. When my sister and I were small children, before we ever heard of Goethe's Faust or Terentius Afer, my mother used to recite a poem every night when she put us to bed. She knew a lot of poems by heart. The one I remember best is "'Tis gone that bright." We liked that one best because it was the longest. It was so long that there was a good chance we would be asleep before it finished. I don't even remember who wrote it. It is a hymn and was probably written by some nineteenth-century hymn-writer.[8] *It begins:*

`Tis gone, that bright and orb`ed blaze,
Fast fading from our wistful gaze...

and it goes on for a long time describing the evening turning into night. Then comes a sudden shift to another theme:

Sun of my soul, my Savior dear,
It is not night if thou be near.
O may no earthborn cloud arise
To hide me from my Savior's eyes.

After that there are many more verses but I don't remember how it ends. When my mother was in a hurry, she would begin with "Sun of my soul" instead of "'Tis gone that bright," and we would be disappointed.

When we were older, another book that my mother liked to quote from was "1066 and All That" by Sellars and Yeatman, two school-teachers who wrote the book as a spoof of English history as it was taught in schools. The book was written in the 1930s and was a best-seller. I only remember the last sentence of the book: "In 1920 America became top nation and history came to a stop."[9] *Before that, of course, England was top nation and that was what history was about.*

STS to FD:

4. As a professor, what ways have you found most effective to spark the hunger for knowledge in others?

FD to STS:

4. I don't know whether I ever sparked the hunger for knowledge in anybody. That is not my main purpose when I am teaching. When I am teaching, I treat the students as grown-ups and try to engage them in discussions. I am not trying to spark their hunger for knowledge. I am trying to get them to think. I assume that if they were not hungry for knowledge they wouldn't come to my classes, and in that case they probably shouldn't be in college in the first place.

I have done two very different kinds of teaching. Recently I have mostly been teaching "Science and" courses, such as "Science and Society," "Science and Literature," "Science and the Environment," to mixed classes of humanities and science majors. Earlier I was teaching mostly technical physics courses to science majors. In the technical courses, the students have to do problems. In the non-technical courses, they have to write essays. For me, the problems and the essays are the most important part of teaching. By forcing the students to solve the problems or write the essays, and then criticizing and discussing their efforts, I am teaching them to think. That is the best way I know to get them to think. Also, to think clearly and write clearly are the most important skills that they will need when they are out in the real world. I don't think this has much to do with sparking their hunger for knowledge.

I have only been teaching university students. For an elementary school teacher, sparking the kids' hunger for knowledge is the main object of the game. Some of the teachers who teach my grandchildren do it very well. I don't know how they do it.

STS to FD:

5. Another book we read this semester was *The Ascent of Man* by Jacob Bronowski.[10] In his final chapter, "The Long Childhood," Bronowski describes John von Neumann ("the cleverest man I ever knew"), who, in Bronowski's view, "was in love with the aristocracy of intellect." Bronowski went on to say that we must be a "democracy of the intellect." Since you probably knew John von Neumann at the Institute, would you care to critique Bronowski's comment on von

Neumann, and of the danger of a too-close relationship between science and the high seats of government and power?

Thank you again Professor Dyson for your generous gift of yourself to our class... Best wishes to you and your family...

FD to STS:

5. I didn't know von Neumann well and never talked with him about his philosophy of life. During the time I knew him, he was working hard against great obstacles to get the first modern computer operating here at the Institute for Advanced Study. The computer was being used for three main purposes,

(1) simulations of hydrogen bombs about which I knew nothing,

(2) simulations of weather aimed at improved weather-prediction,

(3) simulations of evolution of living organisms in an artificial universe.

There was a lively group of young meteorologists working on problem (2). Most of them came from Norway, which has always been a great country for meteorology since it has atrocious weather and depends on fishing to stay alive. I enjoyed talking with them and hearing their stories. They all had a great respect for von Neumann.

...I knew Bronowski well, better than I knew von Neumann.... Both of them were fundamentally decent people trying to do good for humanity,... I happen to agree with von Neumann that it is not wrong, and it is often part of our duty to society, for scientists to talk with generals and admirals. Generals and admirals are isolated from the real world, and we should talk with them whenever we get the chance, just as Jesus talked with publicans and sinners... If von Neumann and Bronowski were running for president (what a delightful prospect that would be!) Bronowski would probably win, but I would vote for von Neumann.

That's all for this year. Happy Christmas and New Year to you and the students! Yours ever, Freeman

As a member of the Jasons, Professor Dyson talks often with the generals and admirals, and keeps them from being isolated. We hope they listen to him.

STS to FD:

29 December 2000

Dear Professor Dyson,

Best wishes to you and your family for a joyous Christmas. Enclosed is a Christmas card from the Fall 2000 STS class. ... Also enclosed are momentos of our field station in Costa Rica....

-STS class

Sample student comments written in the greeting card included

"Thank you for sharing such very real insight. You have given so many so much! Holiday Greetings to you and all your family," Ashley Gill

"I appreciate not only your devotion to education, but particularly your dedication to students, even those at a university like ours, far removed from yourself."
Chad Collins

Wishing you a merry Christmas
bright with celebration,
wrapped in favorite memories,
and warmed by the love
of family and friends.

8 Overcoming a Conflict Between Truth and Loyalty

A soft answer turneth away wrath. —Proverbs 15:1

There was a small amount of fallout from Professor Dyson being awarded the Templeton Prize for Progress in Religion. One physicist published a letter in *Physics Today* claiming that the Prize dupes scientists into becoming spokespersons for conservative religious causes.[1] Professor Dyson responded to the criticism with patient eloquence:[2]

The Editor, Physics Today
American Center for Physics
One Physics Ellipse
College Park, MD, 20740
Letter to the editor:

As a Templeton prize winner I wish to reply to the letter of Mark Friesel in your February 1 issue not to defend myself, but to defend the Templeton Foundation against the accusation that the prize "is a bribe that has successfully lured more than one well-known scientist into becoming a spokesman for the right-wing religious cause." The Templeton Foundation is well aware of the harm that has been done in the past by religious intolerance and fanatical belief. The Foundation does not discriminate between Christianity and other religions. Templeton prizes have been awarded to Jews, Moslems, Hindus and Buddhists as well as to Christians. The main purpose of the Foundation is to change religion from a regressive to a progressive force in the modern world. This is a goal that all scientists, whether or not they are religious believers, can share.

Yours sincerely,
Freeman Dyson

Professor Dyson graciously copied me in his response to *Physics Today*. At that time he asked about a picture of an endearing little girl whose image appeared on a note I had sent to him from Costa Rica some weeks before (Fig. 8.1).

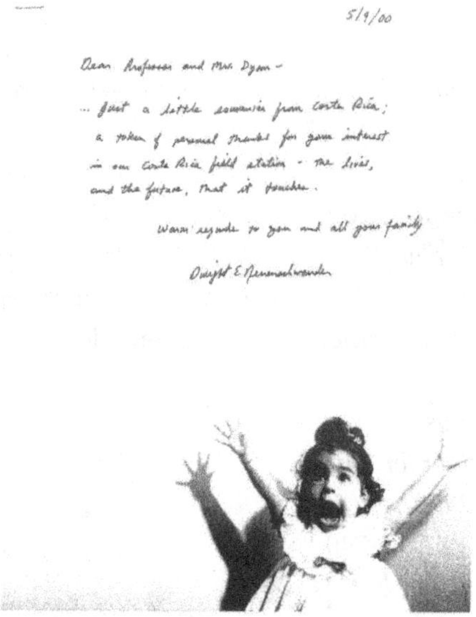

5/9/00

Dear Professor and Mrs. Dyson –

... just a little souvenir from Costa Rica; a token of personal thanks for your interest in our Costa Rica field station – the lives, and the future, that it touches.

Warm regards to you and all your family

Fig. 8.1. Note sent to Professor Dyson sent during a student astrobiology expedition in Costa Rica.

DN to FD:

8 February 2001

Dear Professor Dyson,

Thank you for copying me in your letter to *Physics Today*... I think it's safe to say that persons in a "right-wing religious cause" will... miss the point of Sir John Templeton's efforts. ...

Your message reminds me that I owe you a response from your note of a few days ago. I wish I knew who the little girl is in the stationary picture. I can't look at her without being tickled. Any little person who has such a joy of life has got to be interesting and great fun to be around. I found the stationary being used to post announcements in our music department here, and I convinced the music secretary to give me several sheets of it.

Last Friday I received from George a copy of his book *Baidarka*[3] plus some plans of his more recent boats.... I showed the students pictures of his boats and the tree house. Does he still have the tree house? Is he still living along the Pacific coast?...

Warm regards, Dwight

FD to DN:

9 February 2001

Dear Dwight, thanks for information about the little girl (but my granddaughters are even cuter). I already got a friendly response from Mark Friedel, so it seems I was successful in turning an enemy into a friend. I am glad you are in touch with George. He lives on the west coast but not in Canada. He moved about ten years ago to Bellingham just south of the border. But his boat-business is now completely computerized, so he won't need your manual skills any more. The tree-house still exists, but he has made it inaccessible so kids can't climb up there and kill themselves. As it is a hundred feet up, the risk of a kid doing something stupid is real.

Yours ever, Freeman

Our next set of questions to Professor Dyson was sent near the end of the Spring 2001 semester.

STS to FD:

1 May 2001

Dear Professor Dyson,

Once again as we near the semester's end, our Science, Technology, and Society class sends greetings to you and your family. We also have some questions, and if you could respond to two or three of them, we would be very grateful...

Before going to the questions, we want to thank you again for your support of our work in Costa Rica. Your generosity helped three students visit our Quetzal Education Research Center in the Costa Rican high-altitude cloud forest. In addition, they visited the Arenal Volcano and the Pacific coast at Manuel Antonio National Park. We have some more momentos of the trip to send to you...

Now to the STS students' questions. These were selected by the class in an open discussion last night. Thank you for any comments you may have in response to our questions.

1. STS to FD: What decision by scientists has most affected the scientific community in a negative way?

FD to STS:

3 May 2001

Dear Dwight,

Thank you for the May Day message and thanks to the students for their list of questions. Please say a special thank-you to Beth Balderas for her personal letter, which I hope to answer by and by.

Two misfortunes fortunately canceled each other out. (1) Your list of questions came too late for me to answer before your deadline of May 7, since my wife and I are supposed to be in Texas with a gathering of high-school students from all over the USA. (2) My wife collapsed suddenly with acute appendicitis and is now in hospital recovering from the surgery. As a result of (2), we canceled the trip to Texas and I have time to answer the questions. As you can imagine, I am a bit frazzled, and the answers may not be very illuminating.

1. It is hard to find examples of scientists doing things that affected the whole scientific community, either for good or for bad. The community is diverse and divided up into a thousand little communities with different interests and different problems. The best example I can think of comes from the "McCarthy era" in the 1950s, when many scientists lost their jobs because they were accused of being Communists. The victims were questioned publicly by Congressional committees in Washington, who demanded that they talk about

(a) their own political activities,

(b) the activities of their friends. The victim then had a choice of two responses. Either

(c) refuse to answer all questions by claiming the protection of the Fifth Amendment to the US Constitution, which says that nobody can be compelled to testify against himself, or

(d) answer freely the questions about their own activities but refuse to answer questions about their friends.

Among my own friends at that time, David Bohm chose (c), Wendell Furry and Chandler Davis chose (d). If you chose (c) you were legally protected. If you chose (d) you might go to jail. But the effects of people choosing (c) were disastrous for the scientific community. To plead the Fifth Amendment was generally understood to mean that you were guilty and had secrets to hide. This played into the hands of the politicians like Joe McCarthy who were scaring the public with scare-stories of Communist conspiracies. What happened to the individual victims? David Bohm came out of it badly. He was indicted for contempt of Congress, tried and found not guilty, but this did not do him any good. He lost his job at Princeton anyway, and had to leave the USA to find a job in Brazil. Chandler Davis went to jail for a year and is now a distinguished professor in Toronto. Wendell Furry was never indicted and continued his life at Harvard without interruption. If everyone had chosen (d) and accepted the risk of going to jail, the damage to the country would have been much less.

Last year we had a similar witch-hunt with the accusations against Wen Ho Lee at Los Alamos.[4] *Wen Ho Lee went to jail for a year and the accusations that he was a spy collapsed. With luck, this witch-hunt may now be coming to an end. The courage and honorable behavior of Wen Ho Lee certainly helped to limit the damage.*

STS to FD:

2. How would you recommend that we educate K-6 students about science?

FD to STS:

2. I don't know much about educating kids about science. My own grandchildren seem to be picking up a lot of science, but I don't know how much comes from school and how much from parents and friends. We are lucky to have a number of excellent museums in this area, and kids probably learn more in museums than they do in school. They love to go on trips to museums. Recently I picked up a crinoid fossil when I was hiking in Arizona. I gave it to one of my grandsons and he said at once, "Oh yes, that was living on the bottom of the sea before there were dinosaurs." I asked him how he knew that, and he said he had seen one in a museum. It is amazing how much they pick up. The main thing to avoid is turning them off science by forcing them to learn a lot of dull stuff in school. If I were in charge of public education, I would campaign for more museums and fewer classrooms.

The next two questions were motivated by the indelible final scene in *Disturbing the Universe*. It will be recalled that in the narrative of the dream, when Professor Dyson and two of his daughters arrive in God's throne room, the wicker throne at the top of the steps appears to be empty. After some time Professor Dyson ascends the steps and finds on the throne a three-month-old infant smiling at him. It is so interesting that student's questions about this scene keep recurring over the years.

STS to FD:

3. In the final chapter of *Disturbing the Universe*, who or what does the baby represent?

FD to STS:

3. The baby at the end of Disturbing the Universe does not represent anything. It is just what I saw in the dream. I do not think it makes sense to ask what it represents. Each of you is free to make your own interpretation of it. When you see a painting of a baby in an art gallery, for some people it is baby Jesus, for others it is just a baby, for others it is a clever arrangement of paint on canvas, and so on. If you want to make a theological interpretation of it, you can say that it says that God is only at the beginning of his work in the universe, and perhaps that is true.

STS to FD:

4. In that final chapter, what were the questions you were going to ask God?

FD to STS:

4. The questions I was going to ask were the usual questions. Why are you so unfair? Why do some people get all the luck? Why are most of us cheats and crooks? Couldn't you have made us better? And so on.

STS to FD:

5. Who is (or was) the most influential person in your life?

FD to STS:

5. My life has been many-sided, and different people are most influential in the various sides of my life. In my life as a scientist, the Russian mathematician Abram Besicovitch, who set my style and my way of solving problems. In my life as a writer, the writer Herbert [H.G.] *Wells, who wrote a first-rate history of mankind*[5] *as well as wonderful novels and a textbook of biology. In my life as a human being, my mother Mildred Dyson, who formed my tastes in literature and also in religion.*

Professor Dyson's mother Mildred appears in multiple responses to our questions about the most influential persons in his life. Professor Dyson loved literature and has been an effective communicator to the general public as well as to specialist colleagues. He fulfilled two distinguished careers, one as a scientist and another as a writer. He quotes a line from his Cambridge mathematics professor G.H. Hardy: "Young men should prove theorems, old men should write books."[6]

After the Sloan Foundation invited him to write his memoir that became *Disturbing the Universe*, Professor Dyson quipped "Life begins at fifty-five, the age at which I published my first book."[7] Although he is a powerful mathematician, his memoir about science and technology consists of stories rather than equations. "The methodology of this book is literary rather than analytical." [*DU* p. 5] Books are clearly an important part of Professor Dyson's life.

STS to FD:

6. What book was the most influential in your life?

FD to STS:

6. The most influential book was probably Eric Bell's Men of Mathematics,[8] *a collection of biographies of mathematicians written in a readable and racy style. A large proportion of the mathematicians of my generation were seduced by Bell's book into becoming mathematicians. The great thing about Bell's book is that the characters are real people with many faults and weaknesses, so the stories ring true. If this collection of jerks and fools could do great mathematics, then so can you.*

STS to FD:

7. What makes you laugh?
Thank you again Professor Dyson for teaching us so much....
Best regards to you and Imme and your extended family.
Warm regards, Spring 2001 STS

FD to STS:

7. What makes me laugh is mostly my grandchildren (Fig. 8.2). *I am lucky to have three of them living close by. Also, when I am sitting at the lunch table with the local astronomers talking shop, we laugh a great deal. In a profession like astronomy, or music or theater, with a large number of prima donnas, whenever a group of experts comes together, they will be making jokes about the antics of the prima donnas.*

That's all for tonight. I still have to visit my wife in the hospital before going to bed. Thanks again for your questions, and all good wishes for your futures.

Yours sincerely, Freeman Dyson

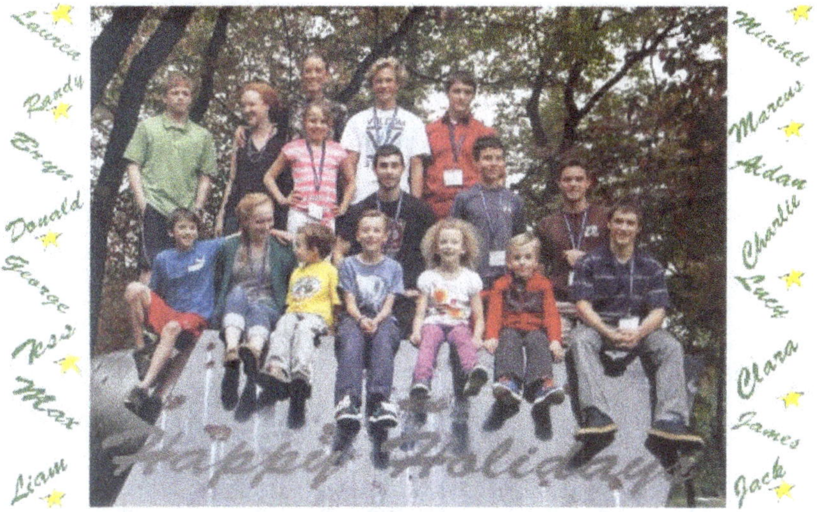

Fig. 8.2. The Dyson grandchildren at the Albert Einstein Memorial, Institute for Advanced Study, Princeton, NJ, September 2013, on the occasion of IAS hosting a 90th *birthday celebration for Professor Dyson. Photo courtesy of the Dyson family.*

DN to FD:

> 4 May 2001
>
> Dear Professor Dyson,
>
> Just a quick note of acknowledgment and thanks for your reply to the questions of our class.
>
> First, we are sorry to hear that Imme fell suddenly ill with acute appendicitis. Our thoughts & prayers are for her and you and your extended family at this trying time.... All the more, then, our reason to thank you for taking time out during these stressful days to answer a bunch of questions. Would that you and Imme made your trip to Texas (our questions could have waited), and I am sorry that the high school students did not get to spend time with you....
>
> Thank you again, and best wishes for Imme and you.
>
> Warm regards, STS class

After the Spring 2001 semester a summer miniterm session commenced. A brief note asked Professor Dyson if he would be "open to questions from this class too, so soon after having questions dumped on you from the regular semester class?"

FD to DN:

23 May 2001

Dear Dwight, thanks for your message which reaches me at Jet Propulsion Laboratory in California. I will be home on May 26 and I look forward to the next batch of questions.... Imme is slowly recovering and hopes to be running again soon. I will be leaving for England on June 15, so it would be helpful to have the questions next week rather than later.

Yours, Freeman

A few days later came the questions from the summer school mini-term session:

STS to FD:

4 June 2001

Dear Professor Dyson,

We hope this week finds you rested from your visit to JPL, and also finds Imme on the mend. Here are a few of the questions from the STS mini-term class....

1. Do you believe that a gray technology that has an awareness of itself could have a sort of soulishness, or in other words, an ability to create relationships with entities around it? We read George's preface to *Darwin among the Machines*,[9] where he talks about trees thinking and engines having souls.

FD to STS:

10 June 2001

Dear Dwight, congratulations for having survived the two weeks of hard labor. That must have been a strenuous, as well as a rewarding, exercise. In my privileged life I never had to teach as much as four hours a day. And still, an average high-school teacher would consider that a light teaching load. It is amazing that so many good people are still teaching in high-schools.

I have a quiet day alone while Imme is in San Diego celebrating our daughter Emily's fortieth birthday. I didn't go for the birthday because I was there two weeks earlier on my way to JPL. As I may have told you before, Emily has a new son Marcus, born on May 8. His brothers aged 4 and 2 seem to be tolerating him well. Emily inherited some good motherhood genes from Imme.

I also have a cheerful postcard from Michael Crabtree [SNU Vice President for Development] *in Costa Rica reporting the official opening of QERC.... I am happy that the project now has official status and formal blessing from the Costa Rica government.*

Now for the students' questions. I am afraid my answers will be disappointing. At 77 my memories are not as sharp as they were when I wrote the book.

1. The question whether a machine can have a soul is one of the fundamental questions that lie at the root of our understanding of our own nature. Of course, nobody knows the answer. We are, after all, machines of a different kind. Does God look for the DNA each time he hands out a soul? Or does he count the neurons and synapses? When my son George writes about trees thinking, and a ship's engine having a soul, he is writing poetry and not science. I do not believe literally that a tree can think or a ship's engine dream. But I find these poetic images full of meaning. There is another world out there, the world of trees and ships, and that other world may well have a mind of its own, or at least the capacity to grow a mind when it gets the chance.

I have a theory about the human soul, which says that the essential feature of our brain that allows the soul to exist is randomness. In every human brain there are about a hundred trillion synapses, connections between neurons, which grow with a high degree of randomness during our first two years. It is the randomness of the synapses that gives us the chance to be creative, to be different from one another, to be individual souls. When God hands out the soul, he looks for that creative randomness. The machines that we build today, whether they are ship's engines or supercomputers, do not have this quality of randomness. Therefore, they do not have souls. When machines are built by us or by one another, with randomness equal to ours built into their structure,

I would not be surprised to find that they also have souls. Our grey technology today is very far from being able to build such machines. But it is possible, as my son George has said, that the world-wide community of machines that we call the internet will develop into a kind of organism with enough randomness to grow a soul. That is a possibility that we should all take seriously. It is not the individual machine that threatens our dignity as human beings, but the world-wide community of machines. We rely on the machines to organize our lives, and we are growing increasingly dependent on them. When they develop a mind of their own, we may be in trouble.

Our second question was motivated by Chapter 3, "The Children's Crusade," in *Disturbing the Universe*. A crucial passage reads,

Bomber Command might have been invented by some mad sociologist as an example to exhibit as clearly as possible the evil aspects of science and technology: the Lancaster, in itself a magnificent flying machine, made into a death trap for the boys who flew it...A bureaucratic accounting system which failed utterly to distinguish between ends and means...A commander in chief who accepted no criticism either from above or from below, never admitted his mistakes and appeared to be as indifferent to the slaughter of his own airmen as we were to the slaughter of German civilians...[*DU* p. 29]

STS to FD:

2. What kinds of emotions did you have to overcome to write the "Children's Crusade"?

FD to STS:

2. *The conflict that I had to overcome when writing the Children's Crusade was the conflict between truth and loyalty. I felt an intense loyalty to the young boys who flew in the bombers and died in thousands in the belief that they were winning the war against Hitler. And I felt an intense shame for having failed to speak the truth during the war, the painful truth that the sacrifice of these*

young lives was in vain, that our losses were out of proportion to the military damage we were doing to Germany. It took me twenty-five years to resolve the conflict. The first time I wrote about the bombing campaign was in 1970, in a piece called "The Sell-out" published in the New Yorker (February 21, 1970) and afterwards reprinted as chapter 10 in my book "Weapons and Hope" in 1984.

Let us interrupt Professor Dyson's letter to recall a passage from *Weapons and Hope* where he shows how the lines between the "good guys" and the "bad guys" can become blurred. His analysis of RAF bombing campaign results made him "sickened by what I knew."

Many times I decided I had a moral obligation to run out into the streets and tell the British people what stupidities were being done in their name. But I never had the courage to do it... After the war ended, I read reports of the trials of men who had been high up in the Eichmann organization. They had sat in their offices, writing memoranda and calculating how to murder people efficiently, just like me. The main difference was that they were sent to jail or hanged as war criminals, while I went free. I felt a certain sympathy for these men. Probably many of them loathed the SS as much as I loathed Bomber Command, but they, too, had not had the courage to speak out. Probably many of them, like me, lived through the whole six years of war without ever seeing a dead human being.[11]

Let us return to Professor Dyson's letter of June 10:

The "Children's Crusade" chapter that you read covers the same ground in a different way. The resolution of the conflict was to speak truth, even if it was twenty-five years too late, and even if it wounded the surviving bomber-pilots deeply. So I wrote the piece and published it, at the invitation of William Shawn who was then the editor of the New Yorker. It was no accident that this happened in 1970, at the same time that a similar conflict of loyalties was raging, between those who had fought and suffered in Vietnam and those who proclaimed that the war in Vietnam was pointless. The number of British who died in the bombing campaign was roughly equal to the number of Americans who died in the war in Vietnam, and the anguish of those who survived the war was also similar.

After the piece was published, I received a number of bitter, angry and tragic letters from old bomber-pilots, just as I expected. Those letters ought to be published some time, after we are all dead. They are eloquent, passionate, denouncing me for insulting the dead. Those bomber-pilots, who saw so many of their friends and comrades shot down in flames, can never believe that they died in vain. To say that they died in vain insults their memory. And who are you, a privileged young kid who sat out the war in comfort at Command Headquarters, to pass judgment on those who fought and died?

I answered those angry letters as best I could, explaining that I was not insulting the dead when I told the truth about the way they died. But of course, none of the old bomber-pilots forgave me. As long as they live they will revile me as a traitor, a smart-aleck who saved his own skin and then slandered the memory of those who did not save theirs.

How instructive it is that Professor Dyson could empathize with and respect the opinions of the old bomber pilots even as they "denounced" him!

STS to FD:

3. Has your career ever been in conflict with your convictions?

FD to STS:

4. My career never came in conflict with my convictions, because my career has always been concerned with abstruse mathematical science that has no human consequences for good or for evil. My official professional career has been teaching and research in the fashionable areas of theoretical physics, quantum fields and fundamental particles, far removed from things we can feel and touch. However, my career left me free to pursue a broad range of other interests outside theoretical physics, and my extra-curricular activities did sometimes lead to moral dilemmas. The most questionable of my extra-curricular activities is to be a member of Jason, a group of scientists who give technical advice to the government in general and to the Defense Department in particular. I am still a member of Jason, and I think I do more good than harm by working for Jason, but I have to weigh my actions carefully. Fortunately, every member of Jason is free to choose which projects to be involved with and which to avoid.

The most difficult time to be a member of Jason was during the Vietnam war. Many people at that time felt that it was wrong to have anything to do with the people who were running the war. Several Jason members resigned as an act of protest against the war. I did not resign, because I thought it was more important to keep talking to the soldiers, to give the soldiers some contact with the world outside the Pentagon. During those years I did not work on the Vietnam war, but I tried to keep open the communication between the protesters outside the Pentagon and the soldiers inside. I don't claim that my efforts to communicate did any good, but at least I did not do any harm.

At the moment I have some conflict arising from my work on climate. I am not really an expert on climate, but I worked on climate at the Oak Ridge National Laboratory for a while and I think I understand the problems quite well. The conflict is the following. The official climate experts have a very dogmatic party line which says that Global Warming is a real and important danger and that we understand it well enough to take action against it. My conviction is that the official experts are wrong, that they rely on unreliable computer models instead of on observations of the real world, and that we should put far more effort into observations before deciding what to do. So my conviction goes directly against the politically correct party line. This is for me a problem. Should I shout or should I keep quiet? On the whole I think it is wise to keep quiet, because I am not a certified expert and any shouting that I do is likely to be ineffective. But I have a bad conscience keeping quiet, because I kept quiet about the bombing campaign during World War Two for similar reasons.

STS to FD:

4. Have you had any more dreams, similar to those described in your book, since *Disturbing the Universe* was published?

Thank you again Professor Dyson. Warm regards to you & your family,
the STS class

FD to STS:

4. Sad to say, I never had any dreams in recent years as vivid as those I wrote about in the book. Perhaps that is a part of growing old. I looked in some more recent papers for interesting dreams, and all I could find was the following item:

January 2, 1995. The most banal conversation ever dreamed. I was in the underworld and met Hitler dressed in his usual Army uniform. I said, "You were not much of a general," and he replied, "You were not much of a general either." I said, "But I was not trying to be one." End of conversation.

Sorry, that is the best I can do. Thank you, students, for another good set of questions. And good wishes to all of you for the summer. And the same to you, Dwight. I look forward to reading the students' written comments when you have time to put them together.

Yours ever, Freeman

STS to FD:

11 June 2001

Dear Professor Dyson,

Thank you once again for your thoughtful replies to our student's questions. I will pass them along to the students through the e-mail or other summer addresses .... Your replies come from the heart, as well as an unusual depth of experience.

Congratulations on the birth of your newest grandson, Marcus. I know you are very proud of all your children and grandchildren. I remember being at the reception for you at St. Albans School, at National Cathedral, before the Templeton Award acceptance speech ceremony last year. I'll never forget how, in the midst of these distinguished people all pressing around you, when some of your grandchildren entered the room you immediately knelt down to have eye-to-eye contact with those wonderful kids, and they ran to you and you welcomed them with open arms. That was wonderful to see....

Thank you again for your responses to our questions....
Warm regards, Dwight

9 A Problem of People's Hearts and Minds

To Mercy, Pity, Peace, and Love
All pray in their distress:
And to these virtues of delight
Return their thankfulness.

For Mercy Pity Peace and Love
Is God our father dear:
And Mercy Pity Peace and Love
Is Man his child and care.

For Mercy has a human heart
Pity, a human face:
And Love, the human form divine,
And Peace, the human dress.

Then every man in every clime,
That prays in his distress,
Prays to the human form divine
Love Mercy Pity Peace.

Then all must love the human form
In heathen, turk, or jew,
Where Mercy, Love and Pity dwell
There God is dwelling too.

—William Blake[1]

Between the previous letter and our next one, the terrorist attacks of September 11, 2001 happened. Three commercial airliners laden with passengers were hijacked and deliberately flown into both towers of the World Trade Center in New York City and into the Pentagon in Washington D.C. On a fourth hijacked airliner that had been turned around and was headed back east, the passengers valiantly fought the hijackers, resulting in their plane crashing into a Pennsylvania field. These attacks were still uppermost on our minds in December when we next wrote to Professor Dyson. We would mention those events in a letter a few months later.

STS to FD:

5 December 2001

Dear Professor Dyson,

We hope you and your family are looking forward to a wonderful time together over the holidays. The events of the past several months have caused us all to remember what is truly important.

If we may trouble you with some questions, the Fall '01 STS class has four for you....

1. In Chapter 20 of *Disturbing the Universe*, "Clades and Clones," you say that a clone is a dead end, while a clade is a promise of immortality. Does that mean that cloning research is directed towards a "dead end?"

FD to STS:

December 8, 2001

Dear Dwight,

Thank you very much for the two packages which arrived safely. I look forward to reading the student journals. Congratulations for playing in the orchestra for the Christmas musical... it's great to be part of the performance, whether you are playing or singing or shifting scenery.

Thanks also for the new message with the four questions. Before trying to answer the questions, I should send greetings to Amber Elder and Beth Balderas who were in Costa Rica last spring and wrote me good letters about their time there. I intended to answer their letters and then forgot. Anyhow, please give them my apologies, and thanks for the letters, and good wishes for the New Year and the rest of their lives. I like the "wisdom of the axe" that Beth wrote about.[2]

So now for your four questions.

1. Is research on cloning a dead end? The answer is no. The word "clone" can be either a noun or a verb, and the noun and the verb have different meanings. As a noun, a clone is a population of animals or plants that all have the same genes. As a verb, to clone means to produce an offspring, bacterium or animal or plant, that has the same genes as the parent. When I said a clone is a dead end, clone is a noun, and I meant that a population with the same genes cannot evolve into something different. But when you ask about cloning, clone is a verb, and cloning is a tool that can be used for many useful and important kinds of research. For example, Ian Wilmut who bred Dolly is interested in using cloning to breed sheep or goats that can manufacture various drugs that are difficult to make by other methods.[3] *Other people are using cloning to understand basic questions about the development of an egg into an adult animal. So research using cloning is certainly not a dead end.*

STS to FD:

2. In Chapter 21 of *DU*, "The Greening of the Galaxy," you remark that "we shall be playing God, but only as local deities and not as lords of the universe." Are you implying that acting as "local deities" is justifiable, and where do you draw the line between being a local deity and trying to be a lord of the universe? You also ask in Chapter 15 ("The Island of Doctor Moreau") whether we can play God and still stay sane, and your answer is no. If we are local deities, are we insane?

FD to STS:

2. You are right in pointing out an inconsistency between the Doctor Moreau chapter and the Greening of the Galaxy chapter of my book. In the Doctor Moreau chapter I am using Wells's powerful myth of Doctor Moreau as an image of a mad scientist playing God with the animals that he has in his power. In the Greening chapter I am imagining a group of people, far away from the rest of human society, who have the right to use genetic engineering to determine their own future. These two images of genetic engineering, one insane and the other sane, are both possible. Where you draw the line is a question for individual judgment. You must look at each situation and decide for yourself whether the use of genetic engineering is justified. I do not try to draw a line that would apply to all situations. In my view, one of the important factors in forming a judgment is whether the use of genetic engineering is imposed by some political authority or freely chosen by the parents. The danger of insanity is generally less, but is not entirely removed, if the parents have freedom of choice.

STS to FD:

3. You say in Chapter 9 that the spirit of the little red schoolhouse is dead. Do you believe that there are still little red schoolhouses? If there are not, that would paint a rather bleak picture of modern science.

FD to STS:

3. Are there still little red school-houses? The answer is emphatically yes. When I wrote in the school-house chapter, "The spirit of the little red school-house is dead," I was referring explicitly to the nuclear power industry. The statement is true for the nuclear power industry, but it is certainly not true for modern science in general, or for modern industry in general. Side by side with the big projects and big companies, there are thousands of little projects and little companies where the spirit of the school-house is alive and well. For example, we have here in Princeton a delightful project run mostly by students, to look for optical flashes in the sky using a one-meter telescope on the Princeton campus. The idea is that alien societies in the sky might be signaling to us with laser pulses rather than with radio messages. The whole project costs only 28 thousand (not million) dollars, to refurbish the telescope and operate the system. We probably will not discover any alien civilizations, but there is a chance we may discover something almost as interesting and unexpected, since nobody has looked for nanosecond pulses in the sky before.

STS to FD:

4. Would you have walked away from Los Alamos during the Manhattan Project?

FD to STS:

4. Would I have walked away from Los Alamos if I had been there during World War Two? Answer, definitely no, since I did not walk away from RAF Bomber Command, and Bomber Command was at least as evil as Los Alamos. In fact I had better reasons to walk away from Bomber Command than Joseph Rotblat had to walk away from Los Alamos. The evil that Bomber Command was doing, killing thousands of our own young men as well as thousands of enemy civilians every month, was much clearer than the evil that Los Alamos promised to do later. I am sure that if I had been at Los Alamos I would have worked on the project enthusiastically just as everyone else did, with the exception of Rotblat, right up to the end. And even now, although I admire and respect Rotblat enormously, I do not altogether agree with him.

That's all for the questions. Thank you for the very interesting information about the student voting. It is interesting that they see genetic engineering as undesirable because it would promote uniformity of human populations, while I see it as dangerous because it would promote too much diversity. Perhaps they are right. But we will never know, unless we let the parents try it and see what they choose to do with it.

Happy Christmas and New Year to you all, yours ever, Freeman

Professor Dyson mentions the one person widely known to have walked away from the Manhattan Project, Joseph Rotblat. In 1944, while at Los Alamos, when it became clear that the Germans had not came close to building an atomic bomb, and that the bomb's postwar purpose would be to intimidate the Soviets, Rotblat walked out as a matter of conscience. In 1955 he was the youngest signer of the Russell-Einstein Manifesto that called on world leaders to eschew hydrogen bombs. Rotblat chaired the press conference where the Manifesto was released, and its favorable reception led to the Pugwash Conferences on Science and World Affairs, held periodically for face-to-face dialog between East and West scientists. Rotblat was a co-founder and secretary-general of the Pugwash Conferences.[4]

The Spring 2002 semester drew to an end as the first anniversary the horrific terrorist attacks of September 11, 2001 approached. Meanwhile, the perpetual Israeli-Palestinian conflict had heated up again. We sent our questions to Professor Dyson on May 7. On this occasion we happened to catch him at a bad time. But he did not brush us off; he never did. We exchanged two more emails that day.

FD to DN:

7 May 2002

Dear Dwight, just to let you know I can't answer your questions just now as I am on the road (at Jet Propulsion Laboratory in California) and will not be back home till May 16. These are good questions and cannot be answered in a hurry. I hope your students will not all be gone by the time I answer them. Sorry I have no time. Apologies to the students.

Yours ever, Freeman

DN to FD:

7 May 2002

No problem... The day our questions become burdensome is the day we back off, and I'm glad you feel free to not try and people-please us. Thanks to email I can keep in touch with students indefinitely, so enjoy your trip & don't worry about our questions until it's convenient for you.

Warm regards, Dwight

On May 20 he answered our questions of May 7. The delay was worth the wait.

STS to FD:

1. STS to FD: In "The Blood of a Poet" [*DU* Ch. 4] you said "A good cause can become bad if we fight for it with means that are indiscriminately murderous. A bad cause can become good if enough people fight for it in a spirit of comradeship and self-sacrifice. In the end it is how you fight, as much as why you fight, that makes your cause good or bad." In light of this truth, would you care to comment on the events of September 11 and its aftermath, and/or the current Israeli/Palestinian tensions?

FD to STS:

20 May 2002

Dear Dwight and students,

I am back from my travels and ready to answer your questions. On May 12 which happened to be Mother's Day, we heard the joyful news from our daughter Mia the Presbyterian minister that we have another grandson. She timed it so cleverly that he was born just after midnight. According to the rules of managed-care bureaucracy, that gave her three nights in the hospital instead of two.

1. What do I think about the September 11 events and the Israeli-Palestinian fighting? I think it is too soon to tell whether the cause for which the September 11 hijackers were willing to die, to get the United States military forces out of Saudi Arabia, was good or evil. In my judgment the cause for which the Palestinian suicide bombers were willing to die, to get the Israeli military forces out of their homeland, was good. In both cases, the fact that the fighters killed so many innocent civilians did harm to their cause. By fighting in such a ruthless way they turned a good cause into evil.

As he often did, after answering our question Professor Dyson added a personal story of insightful relevance. His extended response to Question 1 took us back to London during the Blitz in 1940. Young Freeman Dyson lived through it.

In this connection I would like to tell you about a vivid and uncomfortable memory from my own younger days. I am sixteen years old, an angry kid lying in bed in London in September 1940. Although I have been brought up as a privileged child in England, I am violently hostile to the British Empire and everything it stands for. I hate London, the citadel of oppression, with its huge buildings sucking the wealth from every corner of the world. Overhead the German bombers are droning. I lie in bed listening to the bombs exploding and the buildings crumbling. What joy to hear, after each explosion, the delicious crunch of buildings falling down, the great British Empire audibly crumbling. The joy far outweighs any fear that my own home might be hit, or any pity for the people dying in the burning buildings. When I see now on television the pictures of the World Trade Center buildings collapsing, I think, how many angry sixteen-year-olds all over the world are feeling the same joy that I felt in 1940. I find it easy to imagine the state of mind of the young men who so resolutely smashed those planes into the buildings. Almost, I could have been one of them myself.

The only wisdom that I can extract from this memory is that the problem of terrorism is not a military problem. It is a problem of people's hearts and minds. Attempts to solve it by military means will only make it worse. I don't pretend to know how to solve it. A good way to start would be for our country to stop telling the rest of the world how to behave. We must learn to live with the world as it is, not as we want it to be. We must treat our enemies with respect, so that we do not appear to be trampling on their cultures and traditions. The ultimate goal must always be, not to destroy our enemies but to convert them into friends.

The reading of this reply aloud in class was followed by a spontaneous and very long silent pause. Would that Professor Dyson's last two sentences were the guiding principle of every nation's policy.

STS to FD:

2. What is your take on stem cell research?

FD to STS:

2. What is my take on stem-cell research? Here I have a simple answer. I think the British law on stem-cell research is reasonable and wise. The British law says that human embryos left over from treatment of patients in fertility clinics may be used freely for research until they are fourteen days old. As soon as they are fourteen days old they must be destroyed. This law was worked out after a long and public discussion involving religious leaders, politicians, scientists and concerned citizens. The fourteen-day limit means that the embryo is still a little bag of undifferentiated cells without anything resembling a human body-plan. These embryos would in any case have been destroyed if they had not been used for research. The law allows a great deal of useful science to be done and is acceptable to all except a minority of religious believers.

In contrast, the US law says that research may be done using cell-lines that were derived from human embryos several years ago but may not be done using new embryos discarded by fertility clinics. This is a stupid law, the result of a political process carried on without much public discussion. The old cell-lines are of dubious quality and certainly not adequate in the long run for the needs of science. The distinction between old cell-lines and new cell-lines makes no sense in terms of religious principles. So the US law impedes science without any logical basis in religious principles.

We still do not know how important the knowledge derived from stem-cells will be. That is the nature of science, that you can never predict which way it will lead. Certainly we cannot claim that stem-cell research will lead directly to cures for human diseases. All we can claim is that understanding the process of development that turns a single cell into a human baby will provide new insight into the causes of hereditary diseases. Understanding the causes may allow us to avoid the human tragedies that hereditary diseases bring with them.

STS to FD:

3. Another voice we are reading this semester is Robert Pirsig's *Zen and the Art of Motorcycle Maintenance.*[5] Among many other things, Pirsig emphasizes the importance of living in the present moment and spends much of the narrative describing the details of what's around him at a given time. It's like he emphasizes "being" besides "doing." How do you link your passion for the future with the here-and-now of the present moment?

FD to STS:

3. *How do I link my passion for the future with Pirsig's emphasis on the present moment? The best answer I can come up with is to say that I admire Pirsig but do not try to be like him. He and I have many different interests. We agree about the importance of tools and of taking care of young people. But he is intensely interested in*

philosophy and I am not. I am intensely interested in the long-range future and he is not. I am glad you are reading his book, which contains a lot of wisdom and is beautifully written. But it would be boring if Pirsig and I agreed about everything.

One of Pirsig's themes says that although we cannot *define* Quality, we *know it* when we see it.

STS to FD:

4. Another question we raise also relates to Pirsig's book. We ask it because you mention *Zen and the Art of Motorcycle Maintenance* in "The Magic City:" [*DU* Ch. 1] How do you define (or describe or characterize) "Quality"?

Thank you, Spring 2002 STS class

FD to STS:

4. I don't try to define quality. If you need to define it, then you don't understand it. Defining quality is a futile exercise, like defining good and evil or beauty and ugliness. I mean by quality a hoe-plough that someone in Idaho sent as a present to my wife, a simple hand-tool that makes it easier and pleasanter to weed a garden. Or a mountain bike that she likes to ride into town to do the shopping. Or a neat piece of writing like Pirsig's "Zen."

That's all I can say in response to your questions. I don't know whether you are aware that Pirsig's son Chris was murdered as soon as he was grown up.[6] *A senseless death on a street in San Francisco. Apparently a random act of madness without any motivation. This must have been a crushing blow for Pirsig. It reminds me how lucky I am to have a son who is nearing fifty and still alive and vigorous. I don't know whether Pirsig is still writing.*[7]

Please say thank you from me to Nicole Graves for her friendly letter.... I enjoyed her letter and wish her a great life as a teacher.

All good wishes for the summer to you and the students. And thanks for sending the questions.

Yours ever, Freeman

10 Not as Bad as Imposing Rules on One's Neighbors

Only understanding for our neighbors, justice in our dealings, and willingness to help our fellow men can give human society permanence and assure security for the individual. —Albert Einstein[1]

In the spring of 2003 the United States invaded Iraq on the order of President George W. Bush. Given the tone of the rhetoric leading up to the invasion, the weakness of the evidence presented to justify it, and the histories of zealots in the Administration who were pushing for it,[2] our STS class saw this to be a war started on false pretenses. We observed how it is easier to get into something than to get out of it, and looked to the future of this war with foreboding. I happened to be with a dozen students in Costa Rica on the day the war began. We were in downtown San José, being tourists in the shops before our flight back to Oklahoma the next morning. In a city square a giant TV screen showed live images of the first US air raids over Baghdad. The expressions on the faces of the Costa Ricans around us seemed to be saying "Have the Americans lost their minds?"

8 May 2003
Dear Professor Dyson,

I hope you and your family are well. Last semester we did not send any questions because we thought it might be well to give you a break from us for a change. But we have not run out of questions, especially in light of recent events. If you have a moment to respond to the following three questions, the class would be much obliged.

1. In light of "The Ethics of Defense," [*DU* Ch. 13] we would appreciate your comments on the USA's strike against Saddam Hussein's regime in Iraq.

FD to STS:

21 May 2003

Dear Dwight,

Good to hear from you again. For the last few days I have been enjoying writing a review of a new biography of Newton. I send you the review as an attachment... The life of Newton is a wonderful example of the unpredictability of human nature. A weird mixture of mathematics, magic, theology and politics, and out of it came modern physics!... Both the review and the book should be published very soon.[3] *Now to answer your questions.*

1. Of course I am strongly opposed to the war in Iraq, both for ethical and political reasons. It goes totally against all the principles I have been preaching, using offensive weapons aggressively against a defenseless adversary. If we were serious about defending ourselves against terrorism, the highest priority would be civil defense at home, and improving our lousy public health system. The war in Iraq has nothing to do with the war on terrorism. I am particularly disgusted with the dishonesty of the Bush regime, telling the American people lies about non-existent weapons of mass destruction in Iraq. I am proud to belong to our local Coalition for Peace Action in Princeton, a group of citizens led by a splendid young Baptist minister called Bob Moore. The Coalition has done a lot of marching and lobbying in the last few months, and we have had some success. Five of the New Jersey representatives in Washington voted against giving Bush the power to make war in Iraq.

As this anthology of the Dyson-STS correspondence is assembled in 2022, hindsight makes it clear that the war in Iraq was indeed started on false pretenses. It also proved more difficult to end than it was to begin.

In the Iraq war the public was told that the US arsenal of self-guided cruise missiles were extremely effective. One danger with any so-called "smart" technology is that as the machines make more of the decisions, human beings may be steered into thinking less. Recalling that he was a member of the Jasons, we were curious about Professor Dyson's thoughts on autonomous bombing that was being advertised as precise by Department of Defense press releases. Was the dubious "Victory through

Air Power" going to be extended to a dubious "Victory through Cruise Missiles"?

STS to FD:

2. Has the development of precision guided bombs changed the situation in regard to air power, as described in "The Children's Crusade"? [*DU* Ch. 3]

FD to STS:

2. The development of precision guided bombs has increased the effectiveness of air power to some extent, but not as much as the public is led to believe. It remains true, as it was true in World War Two, that precision bombing works well when there is no serious defense and not otherwise. In the final months of World War Two, the RAF worked out a splendidly effective way of doing precision bombing at night, and we destroyed all the remaining oil refineries in Germany in a few nights. The method was to have a "Master Bomber" flying low over the target in a small Mosquito airplane and laying flares precisely on the target. While the heavy bombers overhead were dropping their bombs on the flares, the master bomber continued to broadcast instructions in plain language, telling them precisely where to aim. With this system, very few bombs were wasted. But it could only work after the German defenses had collapsed. The same situation holds today, as we saw in Iraq. If the Iraqi defenses had been serious, the bomber airplanes could not have loitered over Iraq, the GPS guidance system would have been jammed, and we would have soon run out of cruise missiles. Unfortunately, the American public has been led to believe that what worked so well in Iraq would also work in other places. If we get into a fight with a country that has serious defenses, as we did in Vietnam, none of these systems of precision bombing will work. The Vietnamese had good surface-to-air missiles supplied by Russia, and they shot down enough B-52 bombers to stop us from bombing precisely. Pray God Mr. Bush will not lead us into another Vietnam.

STS to FD:

3. Against the background of "Clades and Clones," [*DU* Ch. 20] we are curious about your thoughts on our country's increasing measures toward monolingualism. We noted in our discussions that many languages are projected to go extinct in this century, and that one can travel abroad and expect to find English almost everywhere.[4]

FD to STS:

3. I find it encouraging that Spanish is gaining ground in this country rather than disappearing. Recently some cousins of Imme from Argentina came to visit. They speak only German and Spanish, and we were anxious when they decided to spend a day in New York on their own. When they came back to Princeton at the end of the day, they said they had had no trouble at all, because everybody in New York speaks Spanish. It is true of course that most of the rare languages become extinct, but the major languages, those that have a homeland with more than a million people, seem to be surviving well. The fact that English is a second language for much of the world does not mean that the local languages are disappearing. Of course, I still deplore the fact that Americans and Brits don't take the trouble, as educated people everywhere else must do, to learn to be fluent in a second language. It would have been much better if we had kept Latin as the second language for international communication, as we did in the time of Newton. But I am not advocating the revival of Latin. It is too late for that.

Now it is time to go home to Imme and supper…

Postscript, DN to FD:

…..As you know… the past two semesters we have used Robert Pirsig's *Zen and the Art of Motorcycle Maintenance*….

Pirsig provides insight on why reading *Disturbing the Universe* makes a better STS resource than the standard STS texts: the latter create an objective distance or detachment between the reader and the STS issues. STS issues become subjects to be analyzed, whereas *Disturbing* makes them experiences to be lived. This summer I would like to send

> you some more personal reflections. *Disturbing the Universe* was very influential on me personally at a critical moment in graduate school.
> Warm regards, DEN on behalf of 45 STS students

In response to the personal postscript, Professor Dyson added a personal comment in return:

> *Finally, I am looking forward to the personal reflections that you promise to send me later. As always, thanks to the students for their questions and all good wishes for the summer. Yours ever, Freeman*

In a private letter of 13 November 2003 I sent the promised reflections where I tried to explain to Professor Dyson why *DU* means so much to me personally. After more than three decades of teaching undergraduates in a small university that belongs to a fairly conservative denomination, I know that many students come from religious homes where their journeys, questions, and struggles have been similar to mine. As the one who gave us the "two windows" metaphor, Professor Dyson would understand.

While studying the Dyson Equations in graduate school, *DU* appeared in our campus bookstore. The timing was fortuitous. I had come to the place, reached sooner or later by every serious student of science, where one must decide what to make of the materialistic outlook expressed so candidly by Steven Weinberg when he wrote "The more the universe seems comprehensible, the more it also seems pointless."[5] If the universe is ultimately nothing more than mindless mechanical motion, then perhaps it *is* pointless.

I grew up in a parsonage, the son and grandson of pastors. Much about my upbringing I deeply appreciate. Much in church teaching forms solid policies for living, such as the Beatitudes and the Ten Commandments, lessons in the parables of Jesus, the wisdom of the Proverbs, the poetry of the Psalms. The sense of community offered by the local congregation offers support and stability. My parents were not fundamentalists; rather, they were fascinated by what I was learning as a physics major, and showed themselves remarkably open to the ideas I brought home. But in their role they had to minister to all kinds of people, including those in the congregation who were fundamentalists.

As I grew up, I became increasingly uneasy with the loud certainty found in much of the grassroots church culture. If the Divine Mind is an objective reality, why doesn't it reveal itself openly so everyone would know? By the time I reached graduate school I had come to the place where I could be a reluctant atheist. That position offers a simple solution to the problem of God by merely dismissing it. On the other hand, my parents and a few others like them lived every day of their lives consistent with their principles. As I grew up, I was always surrounded by outstanding laypersons in the churches my father pastored, people whose generosity and humility and kindness and selflessness were genuine. Furthermore, unlike pastors and missionaries, they were not professional Christians who were obliged to adhere to a formal creed. The shaping of their integrity by their beliefs was observable. These people's lives were guided by their faith because they believed their faith expressed truth. I have always respected my parents and these distinctive laypersons for their faithfulness to their principles. To them, God was more than a concept. To them, God was a person.

The interpretation that these observations reflect reality about a Divine Personality would make sense if the underlying faith assumptions were known to be objectively true. But by themselves these observations do not prove the assumptions true. So I always felt awkward at being thrust into situations where, as a pastor's son, I was presumed to hold a particular ideology. When I started attending public universities I found that my professors who made no professions of religious faith nevertheless exhibited the values of justice and fairness and grace and kindness and generosity that I saw in the fine church laymen among whom I grew up.

The recurring theme in all these people's lives, inside and outside the church, seemed to be that the individual had taken Self out of the center of the universe. Their life instead revolved around some higher principle, which could be expressed in various but essentially equivalent representations. In my studies of physics and mathematics, I learned that there are many representations of principles and operations. For example, one can do quantum electrodynamics in the formalism of Schwinger and Tomonaga, or with Feynman diagrams. The point is, one representation is not "true" and the other "false." Rather, as Professor Dyson first showed, these different representations of quantum electrodynamics are equivalent.[6] Perhaps the yearnings for meaning and purpose that reside within human beings also finds expression in multiple but fundamentally equivalent representations.

As children we were told by well-meaning Sunday School teachers that we should not doubt. But I had honest doubts, and they could not be dismissed. Doubt and faith go together. Faith is a *choice* about which principle to act upon when one cannot *know.* To have doubt but not face up to it would be dishonest. How could one's relationship with a Divine Mind be based on dishonesty? But the objective existence of God, or a Cosmic Mind, or the Great Spirit remains the great question. Could there be some ultimate reality or dimensions beyond those which are accessible to the human senses, that lie beyond comprehension by the human mind? To not recognize that possibility seems unjustifiably arrogant, contrary even to the spirit of science itself which requires one to be aware of the limitations in one's data and assumptions.

Thus Professor Dyson's closing two chapters of *DU* hit me like a ton of bricks during my graduate school days. His "deeper unifying hypothesis" that he articulated so beautifully in "The Argument from Design" [*DU* Ch. 23] shows that it is not intellectually irresponsible to suppose that one can take seriously the *hypothesis* of a Cosmic Mind. Although we cannot prove, with evidence that will satisfy all reasonable observers, that a Cosmic Mind *does* objectively exist, it is not irresponsible to allow the possibility that a Cosmic Mind *might* exist. In this passage of *DU* Professor Dyson made explicit the distinction between the kinds of questions that belong to religion and those that belong to science. Much of our culture's trouble with loud certainty comes from bulldozing over that distinction. "In the last three chapters of *Disturbing the Universe* you articulated what I had been thinking and feeling for a long time but could not put into words. I thank you for that."[7]

The letter continued,

> The symbol I find the most meaningful in the Christian mythos is the Christ Child of Christmas.... Many people say that the Christmas mythos is secondary, that Easter is what it's all about.... Death is not the worst thing that can happen to us; that is not the fate from which we need redemption.

> Loneliness and meaninglessness are far worse. Viktor Frankl was right when he said that the strength of his love went beyond the physical presence, or even the continued existence, of the beloved....[8]
>
> So at that moment in graduate school, I did not throw overboard the concept that there may be, perhaps, a Cosmic Mind, a source of love and meaning larger than ourselves. The fact that something exists rather than nothing, and we ponder its meaning, seems suggestive that there *could* be a Mind behind it all. That, of course, "is a question that belongs to religion and not to science." [*DU* p. 252]...
>
> Do you know what I see in most students' faces when we finish our discussions about science and religion? *Relief* that they don't have to choose between them, *relief* that science and religion may be seen as complementary and not contradictory.

The letter then moved on to less existential topics. I include three paragraphs of them here because Professor Dyson's responses to these sidebars say much about his character and his approach to decisions. My embarrassingly verbose letter continued with a recollection of that vivid moment in the St. Albans School with his grandchildren rushing in, where they received their grandfather's full attention—

> ...I hope you don't mind that I relate this scene to my students. They break into smiles. Your priorities are appreciated here.
>
> At the Templeton ceremony I met your neighbor as I was walking into the Cathedral. He invited me to sit with him and his wife. He gave me his card which I unfortunately misplaced... He told me what a good friend and neighbor you are, and also mentioned that you could be "fierce" when it came to sticking up for science. I wanted to send them a thank-you card...
>
> Schweber's book[9] mentions that you rode about England on a motorcycle. Also you mention in *Disturbing the Universe* that your father rode a motorcycle. [*DU* p. 105] As a

motorcyclist myself since high school, I find this delightful.... So I am wondering if you rode a Norton or BSA or Triumph during your rides in England, and if you rode when you came to America.

In one of your letters to my class you were kind enough to include copies of various articles and speeches. I would like to reciprocate....

Thank you for sharing yourself with me and with my students.

Warm regards, Dwight

A person of enormous patience, Professor Dyson took the time to respond to this uninvited tome from a floundering soul. Not only did he listen, but he shared some personal news of his own:

18 November 2003

Dear Dwight,

Thank you for the long letter and the package of papers..., I loved especially the Rattlesnake University and the Doubting Thomas.[10] *The Rattlesnake University is the best written, the Thomas is the most thoughtful. Do you know the Gospel of Thomas, one of the gospels that were excluded from the Bible by the enforcers of orthodoxy in the fourth century? Elaine Pagels has written a lot about it.*[11] *Here is one of the verses:*

Jesus said to His disciples, "Compare me to someone and tell Me whom I am like." Simon Peter said to him, "You are like a righteous angel." Matthew said to Him, "You are like a wise philosopher." Thomas said to him, "Master, my mouth is wholly incapable of saying whom You are like."

I would give the same answer as Thomas.

I write quickly now to answer your questions before I go to Washington. Yes, I will be happy to have questions from your class and do my best to answer them. I always enjoy hearing from the students and seeing their thoughts. I liked especially the remark in your letter that you respond more to Christmas than to Easter. I believe this is true of the majority of Christians, but they do not like to admit it. Certainly it is true of me. I am delighted to hear in detail how the students respond to different chapters of Disturbing....

Thank you for your warm remarks about Mia and Kevin and their kids. They now have another boy Aidan, born in May 2002. Did I tell you that we had a grand family reunion in San Diego in August? This was to celebrate my eightieth birthday which is actually in December, but it was easier to get the family together in August. All the six children and twelve grandchildren were together for a week. The mothers and fathers took turns cooking great meals for twenty-four people and organizing expeditions to interesting places. The children all enjoyed this chance to get to know their cousins, and for me it was a fore-taste of heaven, to spend a week with nothing to do but carry babies around. It happened also to be Mia's fortieth birthday and we had a party for her too.

You ask who was the neighbor who sat with you at the Cathedral in Washington.... Several of our Princeton friends were there but no immediate neighbors. It might have been Wallace Alston, who is director of the Center of Theological Inquiry, an offshoot of the Princeton Seminary. Before that he was pastor of the Nassau Presbyterian Church, so we have known him well for many years. Just two weeks ago my wife and I had supper with him and I gave a lecture at the Center... After the lecture we had a session with the local theologians. One of them, a man for whom I have a deep respect, attacked me strongly. He said I am an "arrogant agnostic," trying to impose my set of rules on God, telling God that He is not allowed to reveal Himself to us even if He wants to. I did not find it necessary to defend myself but was glad to let him have the last word. Maybe he is right and I am an arrogant agnostic. I can live with that. Maybe I impose my rules on God, but that is not so bad as imposing them on my neighbors.

Yes, the motorbike I rode in England was a Triumph. ...I sold it when I came to America and never rode again. In America I soon had a wife and kids and staying alive became more important. I am amazed that you say you have been riding ever since high-school and are still alive. I remember in my motorbike years I collected statistics of my motorbike friends and the average expectation of life was about two years. Of course we were then young and reckless and nobody thought of wearing helmets.

Warm greetings as always to the students,

Yours ever, Freeman

I few days later an email postscript arrived:

FD to DN:

26 November 2003

Dear Dwight,

In my response to your last message, I forgot to say that I did read all the papers you sent me last week while I was in Washington. Thank you again for sending them... The two that I found the most moving are the Conversations with Ghosts[12] *and the tribute to your mother.*[13] *That is also a tribute to your father, the real hero of the story.... Recently I heard one of our friends talking badly about her mother who also has Alzheimer's, and I felt very sad. I felt like giving her a copy of your talk. Yours ever, Freeman*

DN to FD:

1 December 2003

Dear Professor Dyson,

Thank you for your gracious comments... Perhaps I needed to write that letter more than you needed to read it. Thank you for your indulgence....

I feel sad for your friend who is struggling in coming to terms with his mother's Alzheimer's. Perhaps their mother is combative; Mom was not.... Whatever the particulars, after a certain point, it's harder for the caregivers than the patient. You are right about my Dad being the real hero of our story. Among his other qualities, I've always admired Dad for always being himself, whatever the circumstances. He's always been one to accept life on its own terms and make the best of it, in good times and bad...

Thank you again....

11 The Mailman is More Important

Out of the current confusion of ideals and confounding career hopes, a calm recognition may yet emerge that productive labor is the foundation of all prosperity. The meta-work of trafficking in the surplus skimmed from other people's work suddenly appears as what it is, and it becomes plausible once again to think the thought, "Let me make myself useful."

—Matthew Crawford, *Shop Class as Soulcraft*[1]

In our class meeting of 23 November 2003, as we discussed candidate questions to send to Professor Dyson, a young lady named Zina Zander raised her hand and suggested "I wonder if there's a question that Professor Dyson would like to ask *us*." What a great idea! Everyone concurred, so Zina's request was forwarded to Professor Dyson ahead of our usual batch of questions. The ongoing US invasion of Iraq inspired his question for us.

FD to STS:

> *11 November 2003*
>
> *Dear Dwight,*
>
> *Happy Thanksgiving to you and the class! I look forward to seeing your questions next week. My wife and I are going up to Maine to celebrate with Mia and her family (now with one more grandson since you saw them).*
>
> *Since I have no time for a careful response to Zina Zander, I ask the class to consider a political rather than a scientific question. How would you react if an army of young Arabs with lethal weapons and knowing nothing of your language and culture were occupying your country and dominating your lives? This question is outside the agenda of your course, but it is an important question and I would like you to think about it. Thank you, Zina, for inviting me to ask it.*

The standard of success in fulfilling the STS course objectives does not depend on how many topics are covered from a pre-decided agenda. Any discussion that helps us think clearly and articulate thoughtfully counts as success. Professor Dyson's correspondence makes us think. At the next class meeting his question became our top priority. Here is a sample of the student responses that we sent to Professor Dyson:

STS to FD:

3 December 2003

Dear Professor Dyson,

Our STS class meets on Tuesday and Thursday, so I brought your question to our collective attention yesterday....

Answering this question was a struggle for us (about 40 students). After presenting the question, I gave the class about five minutes for discussions between neighbors. This was followed by opening the floor to the entire class. Here are samples of their responses... volunteered from the floor....

"I would be scared, as are the Iraquis."

"Out of fear, I would hide in the basement."

"If they were liberating us from a dictator I would feel different than if they invaded to take away our freedoms."

"But even if I lived under a brutal dictator, these guys are still taking over my home. Who are they, to take over my home?"

"If they are overthrowing our government, we would not accept that."

"I would resent them telling me how to live."

> "Be like Gandhi. More terrorism is not the answer."

> "We assume that everyone thinks like us, we deconstruct the Iraqis as not really human beings…."

As the class ended I invited the students to write further comments on slips of paper and leave them on the desk. Here are some representative ones:

> "If 'young Arabs with lethal weapons' were occupying the US and dominating our lives we would see it as an invading force coming to conquer us. Many of the Iraqis feel just that way." -Becky Seville

> "Seeing how our country was so united and were very much a 'brotherhood' after 9/11/2001, I would like to think that if a group of people came into our country to overthrow our government, that we would all unite and take up arms to protect the freedom we are so blessed to have. There is no way that I could hide out in a basement if people were threatening my country, my friends, my family, and my freedom. This freedom is worth dying for and I would if it came to that." -Nathan Clark

Your question generated about 30 minutes of very interesting discussion…. Many of the students had apparently not imagined a role reversal between themselves and the Iraqis.

… We were disturbed that, as a democracy, we initiated an attack on another country without being attacked by them first…. The case for invading Iraq as self-defense was weak, and whether or not Iraq was a terrorist state before, it's becoming one now…

We also came up with about nine questions which I will edit down to about four or five, and send them to you in another message.

Thank you Professor Dyson for your willingness to engage our class in these discussions. It's a rare privilege and we appreciate it.

Warm regards, STS

A few days later we were ready with our questions for Professor Dyson.

STS to FD:

8 December 2003

Dear Professor Dyson,

Thank you once again for your patient willingness to consider our questions. We have narrowed the list to "four plus one...." The "plus one" you will recognize as the question you presented to us.

1. If nuclear chain reactions had not been possible (e.g., had fission released no more than one neutron per nucleus), so that nuclear weapons and reactors were not possible, what do you envision Freddy de Hoffman (given the same resources) might have had you and the others doing in the "Little Red Schoolhouse" at that time in San Diego? [*DU* Ch. 9] In other words, in that setting, with that assembly of talent and resources, what other problem might have captured your imaginations at that time? What would such a group turn its attention to today?

FD to STS:

9 December 2003

Dear Dwight and students,

Thank you for the new set of questions. I congratulate you on finding fresh questions which make me think before I can answer them. Also thank you for adding your names and majors to the message, so I know a little bit about who you are. I give you my answers quickly so you have them before the end of term.

1. This is a very interesting question that I never heard anybody ask before. If fission chain reactions had not been possible, the Manhattan Project would not have existed, and physicists would not have become the politically important elite that they were after

World War 2. Physics would still have been exciting for scientists, but not for generals and politicians.

During WW2 there were serious programs to develop biological weapons in America, Britain and the Soviet Union. The Soviet Union used a tularemia weapon with some success in the battle of Stalingrad in 1942. The USA and Britain had stockpiles of anthrax bombs. My guess is that if nuclear weapons had not existed we would have had a Cold War arms race in the 1950s based on biological weapons, and biologists would have replaced physicists as the politically important scientists. Los Alamos would have been a biological warfare lab, Freddy de Hoffman would have been working at Los Alamos as a biologist, and General Atomic would have been a biotechnology company. (In real life, Freddy did switch to biology and became director of the Salk Institute in San Diego after he retired from General Atomic.) So the little red schoolhouse would have been filled with biologists exploring the new world of biotechnology opened up by the Watson-Crick discovery of the double helix structure of DNA in 1953. I would probably have been happy to switch to biology if Freddy had invited me to work in the schoolhouse in 1956. We would probably have spent the summer working on the genetic code and the possible ways of sequencing DNA. We might have invented ways of using DNA to diagnose genetic diseases and synthesize new drugs. If we had been very clever we might have invented the basic tricks of biotechnology which actually arrived twenty years later, the polymerase chain reaction for multiplying DNA and the gene-splicing technique for moving genes around from one creature to another. The polymerase chain reaction is to biology as the neutron chain reaction is to physics, similar in basic concept and similar in power.

If we had a similar group in the little red schoolhouse today, I imagine it would be looking into applications of nanotechnology, which is now the fashionable new technology as nuclear technology was in 1956. Nanotechnology is the art of constructing structures and devices on an ultramicroscopic scale. One idea I would like to work on would be to build a machine that would attach itself to a molecule of DNA and read out the sequence as it walks along the molecule. If we could do that, we might reduce the cost and increase the speed of sequencing by a factor of a thousand. All kinds of medical applications of sequencing would then become affordable. Another application of nanotechnology might be a molecular imaging device that would wrap itself around an unknown molecule and read out an accurate picture of its shape. I leave it to your imaginations to think of other inventions we might make and other ideas we might explore. That is enough about question 1.

By the way, I recently received some E-mail from Gordon Permann, a volunteer helper at the Barnard School in San Diego, which was the original little red school-house. The school has survived some good times and bad times. Permann says, "Over the summer we rounded up thirty Navy volunteers who painted and scrubbed the place back into some semblance of order, clearing out twenty years of neglect." So the school is still functioning and hoping for better times to come. Permann has written a history of the school since its beginning as a Navy school in 1943.[2]

STS to FD:

2. Who is your favorite poet?

FD to STS:

2. My favorite poet is still William Blake. I won't try to explain why I like Blake so much. Poetry is like music. You feel it but you can't explain it. Blake had strong words to say, usually rebellious, about many different subjects. In Disturbing the Universe there is only one quote from Blake, "Drive your cart and your plow over the bones of the dead" (page 157), which came into my head after Dover Sharp was murdered. In another of my books, Infinite in All Directions, there is more about Blake and several longer quotes (pages 131–134). There I explain why I find Blake to be a kindred spirit. The quotes come mainly from his poem, "America, a Prophecy," written in 1793.

STS to FD:

3. Scientific work often takes one away from home and family. If you had to choose between being with family and

being away from them to cure cancer, how would you find the balance?

FD to STS:

3. The order of priorities in my life has always been, family first, friends second, and work third. This was easy for me because my work was more like a hobby, to be put aside when more serious problems arose. My work was interesting and challenging but not really important to anybody except me. Now you ask what the priorities would be if my work was curing cancer. My answer is, it all depends. That is not a very satisfactory answer, but I think it is the right answer. If the family is running smoothly and the work is at a crucial turning-point, I would leave the family and take care of the work. If the work is running smoothly and the kids are sick and the wife is exhausted, I would leave the work and take care of the family. If both the work and the family are in a crisis simultaneously, I would divide the time between them as best I could. The point is, you have to use common-sense in making such decisions. Judge each situation as it happens, and do not try to follow inflexible rules.

STS to FD:

4. You have worked inside the government as well as outside, such as serving on the Arms Control and Disarmament Agency and testifying before Congress. What changes over the years have you seen from the inside, in how the government presents itself to the public?

FD to STS:

4. I have been working for government as an advisor for almost fifty years, and the main thing that strikes me is how little things have changed over that long time. The politicians at the top announce big changes, but the people who do the daily work in the government departments follow the routines and change as little as possible. On the whole I think this is good. Most of the things the government does are providing services to the public, and reliability is more important than being up-to-date with the latest technology. A good example of this is the monitoring of the environment by the National Oceanic and Atmospheric Administration and various other agencies. Important things to measure when you are studying climate are the sea-surface temperature and salinity of the oceans. You measure sea-surface temperature and salinity by collecting water in a bucket as your ship passes by. It is important never to change the size and shape of the bucket, so that the measurements made this year can be reliably compared with the measurements made ten years ago or fifty years ago. If you change from a metal to a plastic bucket, the measurements will be affected and the comparison is no longer reliable. There is a noticeable difference between the present Bush administration and earlier administrations. The politicians at the top of this administration have less respect for technical competence than any other administration I have known. They seem to take advice only from their cronies and not from outside experts who know a lot more. In spite of the high level of incompetence and ignorance at the top, the government still functions quite well. Underneath the top political level, technical competence is still respected. Needless to say, when I am working for the government, I only talk with people below the top level.

You ask how the government presents itself to the public. I am not sure what this means. It might mean George Bush making speeches on television. Or it might mean the mailman delivering the mail every day, in spite of snowstorms and blizzards. In my opinion, the mailman is more important. And he has not changed much over the years.

STS to FD:

+1. We were also wondering how you would react "if an army of young Arabs with lethal weapons and knowing nothing of our language and culture were occupying our country and dominating our lives."

We wish a happy Christmas to you and your family. We also wish you a happy 80th Birthday.... Our class is planning something for you to help celebrate that event.... We thank you for sharing your life with us through *Disturbing the*

Universe, your other speeches and writings, and your correspondence.

The STS Class, Fall Semester 2003

FD to STS:

Thank you for giving me a chance to answer this question. Thank you also for your answers which Dwight sent to me a few days ago. Your answers were thoughtful and reasonable. As many of you said, we cannot give a definite answer to the question until we know more about the circumstances. Here is my answer for what it is worth.

I assume that there would be two kinds of opposition to the invaders, as there was in occupied countries in Europe during World War 2. There would be young people fighting with guns, and the mass of the population resisting passively with silent non-cooperation. I am too old to be a fighter myself. I would help the fighters as much as I could, and otherwise I would be a passive resister. I would do as much as possible to make the invaders miserable, so that the invaders' government would find it difficult to replace them when they went home.

Happy Christmas and New Year to all of you!

Yours ever, Freeman Dyson

12 The Family Next Door

Digital connections and the sociable robot may offer the illusion of companionship without the demands of friendship. Our networked life allows to hide from each other, even as we are tethered to each other. We'd rather text than talk.

—Sherry Turkle, *Alone Together*[1]

FD to STS:

10 January 2004

Dear Dwight,

Happy New Year to you and the students! And thanks to all of you for my birthday gifts, the majestic birthday card with your heart-warming good wishes on it, and the package of one-page abstracts [student's weekly essays] *that you wrote week by week during the term. I sat down last night and read the abstracts from beginning to end and found them delightful. Such a wonderful variety of personal experiences and personal responses to my little stories. I hope some of your students will go on to become writers. I won't single out any particular names, because that would be unfair to the others, but several of them have a gift for writing. And, more important than having a gift, they have something to write about. I hope you will encourage them to keep on writing. No matter what professions they choose, a habit of writing letters and stories will enrich their lives and give meaning to their experiences. And it may happen that writing becomes a big part of their lives.*

Last May I was giving a public lecture in Portland Oregon, and I met Jean Auel who happens to live there. She is the lady who wrote "The Clan of the Cave Bear,"[2] *the first of a series of books about people living in the ice age 35,000 years ago. She began writing because she was a stay-at-home mom and needed some extra income. And to her amazement, her books sold by the millions and made her an international celebrity. One or two of your students might do that too... Thanks again for sharing their thoughts with me.*

Yours ever, Freeman

STS to FD:

13 January 2004

Dear Professor Dyson,

I am glad that you received our card (selected by the students themselves) and the student's abstracts. Indeed, I will pass along to them your greetings and thanks. May I do that as a letter in the campus newspaper? The editor took STS in the fall semester.

Right now (last week and this week) we are doing an STS section in mini-term... Today I presented the three scenarios that came up in your Gustavus Adolphus seminar on genetic technology... Out of a class of 39 students, the class voted 7 for option 1, 31 for option 2, and 1 person for option 3. When I then revealed how your class voted they were surprised, but agreed with your assessment that option 1 represents freedom; option 2, tradition; and option 3, fairness....

I keep finding references to your work; the *Gravitation* book of Misner, Thorne, & Wheeler contains several.[3] I am continually amazed at the range of topics to which you have turned your mind....

Is there an obituary of Edward Teller that you know of and would like to see published to an audience of physics major alumni? I visited West Point in the early 90's, and my hosts said, "We had Teller here a couple of months ago, and guess what—he's very nice!" That they felt it necessary to talk about their surprise at Teller's human kindness suggests that he was never able to entirely shake off the stigma that was attached to him after Robert Oppenheimer's security clearance hearing. I never met Teller personally but it seems to me that the stigma attached to him was unjust. The significance of your accepting him as your friend so soon after the Oppenheimer trial [*DU* Ch. 8] is a point that I try to impress on students.... your experiences, and your human glimpses of others that we know only from textbooks, helps remove the abstract detachment that comes across from textbooks. The physics community is not very large, and the great discoveries we read about in our textbooks are not far removed from today's students... For example, I'm still amazed that some of the professors I had as a student had met Einstein personally...

Happy New Year. I'm glad your family had a great trip to California. I doubt that the trip included a visit to George's tree house...

Warm regards,
Dwight

22 January 2004
Dear Dwight, thank you for another long message which I found waiting when I returned from California yesterday. I guess your mini-term is now over and it is too late to say Hi to the students. You are welcome to relay any of my remarks as a letter to the campus newspaper...

You ask whether I know of an obituary of Teller... I send you as an attachment an appraisal of Teller that I wrote myself. It is not an obituary but a review of his memoirs, written shortly before his death. It was published in American Journal of Physics... I agree with your remarks about Teller, and I think my review describes him honestly and fairly....[4]
As always, thanks to you and love to the students!
Yours, Freeman

Here we offer excerpts of Professor Dyson's review of *Memoirs: A Twentieth-Century Journey in Science and Politics,* a biography of Edward Teller, written by Teller with Judith Shoolery.

...The second half of this book contains a detailed account of Teller's involvement with weaponry, first at Columbia, then in turn at Chicago, Los Alamos and Livermore, and finally at Stanford. One might expect the narrative in this part of the book to become more political and less personal. But here too, even when

Teller is most heavily engaged in political battles, he portrays his opponents as human beings and describes their concerns fairly. There is sadness in his account but no bitterness.... Throughout his struggles he maintains his talent for friendship. Leo Szilard, who disagreed violently with Teller about almost everything, remained one of his closest friends.

The worst period of Teller's life began in 1954 when he testified against Oppenheimer in the hearing conducted by the Atomic Energy Commission to decide whether Oppenheimer was a security risk.... One result of Teller's testimony was that a large number of his friends ceased to be friends.... Oppenheimer and Teller both suffered grievously from the quarrel, but the damage to Teller was greater...

...Another illuminating passage is a quote from a letter written in 1939 by Merle Tuve, a senior physicist who knew Teller during his years at George Washington University. Somebody at the University of Chicago had asked Tuve for an appraisal of Teller. Tuve replied, "If you want a genius for your staff, don't take Teller, get Gamow. But geniuses are a dime a dozen. Teller is something much better. He helps everybody. He works on everybody's problem...." That was the Teller I knew when I worked with him for three months in 1956 on the design of a safe nuclear reactor.... There was of course another Teller, the Teller who worked crazily for unpopular causes such as hydrogen bombs and missile defense, and who fought furiously for the causes that he believed in. This book gives us a fair portrait of both Tellers, the Teller who gave generous help to young scientists and the Teller who quarreled vehemently with older scientists. Those who disagreed with him did him a grave injustice when they tried to turn him into a demon.

STS to FD:

May 6, 2004

Dear Professor Dyson,

The students in our spring 2004 semester section of our "Science, Technology, and Society" class have enjoyed reading *Disturbing the Universe* and discussing the issues it raises.... If you have a few moments to answer some questions, we would be much obliged...

1. You clearly developed a love for science at a young age, and experienced early its "three beautiful faces" as described in your AAPT speech.[5] When and under what circumstances did you begin to see the "three ugly faces" of science?

FD to STS:

6 May 2004

Dear Dwight,

Thanks to you and the class for a good set of questions. Our son George is here for a week and also sends you his greetings. It is a long time since he lived in the tree-house. Now he is officially here as a historian writing a history of the von Neumann computer project.[6] *He is also a single dad with a beautiful and temperamental teenage daughter. As Shakespeare said, each man in his time plays many parts. Here are some answers to your questions.*

1. The three ugly faces of science are (a) a rigid and authoritarian discipline, (b) tied to mercenary and utilitarian ends, and (c) tainted by its association with weapons of mass murder. When did I first see the three ugly faces? I think (a) came when I was about twelve years old and forced to slog through endless boring exercises in geometry and algebra. Probably (b) came from reading the novels of H.G. Wells, especially "Tonobungay" and "The Island of Doctor Moreau,"[7] *when I was a teen-ager. The final chapter of "Tonobungay" is especially memorable, with the title "Night and the Open Sea." The narrator of the story, after the commercial empire of Tonobungay has totally collapsed, escapes on his high-speed warship ready to sell to the next highest bidder. He stands on the deck of his warship, watching the lights on the shore fade into the distance, his ship swiftly heading into the darkness. Certainly (c) came from my family's memories of World War One, which was a chemists' war, fought with poison gases as well as chemical high explosives. When I was a teenager I also read Aldous Huxley's "Brave New World" which begins with a war fought with anthrax bombs.*[8] *In those days we expected World War Two to be a biologists' war fought with biological weapons. We knew about anthrax. Anthrax was another gift that we owed to science.*

STS to FD:

2. How has your experience in science influenced your perspective on human rights?

FD to STS:

4. How has my experience in science influenced my perspective on human rights? So far as I can remember, I don't see any connection between science and human rights. I remember when I was visiting Berkeley in California in 1948, there was a lecture on civil disobedience by a young man I had never heard of. His name was Martin Luther King. I went to hear the talk and wrote home to my mother: "This is a man I would be glad to go to jail for." Later I heard him again in Washington when he spoke at the Lincoln memorial and said "I have a dream. [*DU* pp. 140–141][9] *This had nothing to do with science. Some of the leaders in the fight for human rights have been scientists, the most notable being Andre Sakharov, but others such as Gandhi and King and Mandela were not. I think I would probably have been more concerned with human rights if I had been a lawyer rather than a scientist.*

From our perspective and observations, Professor Dyson has always been deeply concerned with human welfare. Human welfare concerns may not be part of his professional job description as a scientist, but his writing career shows human rights and human welfare to be an integral part of who he is as a human being.

STS to FD:

3. You demonstrated how "a good cause can become bad if we fight for it with means that are indiscriminately murderous. A bad cause can become good if enough people fight for it in a spirit of comradeship and self-sacrifice." [*DU* p. 41] The attacks of September 11, 2001, were unquestionably bad, but what was the good cause whose fate it was to be made bad by those attacks? In other words, what was the message that we

should be listening to from responsible people in the Arab world? And has our fighting back by taking the war to them turned a good cause into a bad one?

FD to STS:

3. This is a difficult question. We still do not know exactly what the September 11 gang had in mind. Perhaps we will never know. We know that Osama Bin Laden resents the existence of American military bases and commercial operations in Saudi Arabia, and is trying to foment violence so as to drive the Americans out of the Arab world. I do not consider this to be a noble cause, but it is understandable. Certainly we played into his hands by invading Iraq. The invasion of Iraq turned a local squabble into a full-scale war and turned our good cause into a bad one. The message from moderate Arabs that we ought to have listened to was: "Give the Palestinians some support in their fight for independence, and stop your unconditional support of Israel."

STS to FD:

4. In "A Distant Mirror" you note, "Technology has caused, and will cause, profound changes in style of life and thought, separating us from our neighbors. All the more precious, then are the bonds of kinship that tie us together." [*DU* p. 173] One of our students wrote in an essay how she saw a family in a shopping mall, and each family member was talking on a cell phone to someone who was not there. Thus while the family was physically together, they were apart.[10] Do you see the "separation" of technology happening on the short time scale as well as long timescales?

Thank you Professor Dyson for sharing your experiences and insights with us.

Warm regards,

DN and the 38 students in the Spring 2004 STS class

PS: We have students going to Costa Rica in May. Some of them have applied for the Dyson Travel Scholarship to subsidize their study and research in the tropical cloud forest and surrounding areas.... Thank you for being our catalyst.

FD to STS:

4. I agree with your student that cell-phones are a wonderful example of the way technology separates us from our neighbors. The picture of the family at the shopping mall is a picture of the way we are all going. I am not yet addicted to cell-phones but I am already addicted to E-mail. I notice that as a result of E-mail I have many more friendships all over the world, and many fewer here at home. Instead of inviting our neighbors over for supper, I sit here in the office writing E-mail to you. I cannot regret my addiction to E-mail. After all, my own large family is scattered over thousands of miles, and it is E-mail that has brought us closer together. But we pay a price for this distant closeness. The price is the family next door that we don't really know. So the answer to your question is yes. Technology is separating us on the short as well as the long time-scale.

That's all for today. Now I end by wishing you all a good and peaceful summer, especially those who are going to Costa Rica...
With thanks and good wishes to all of you,
Yours ever, Freeman

In 2004 the Sigma Pi Sigma physics honor society was making final plans for its quadrennial convention to be held in 2005 at the University of New Mexico. A guided tour of the Trinity Test Site was arranged for the participants. Given the history of this region, the conference theme was "Science and Ethics." We asked Professor Dyson if he might be available as a plenary speaker. As it turned out, he was otherwise engaged on those dates. But he offered something in return.

FD to DN:

30 August 2004

Dear Dwight, thank you for the personal note on the invitation to the Sigma Pi Sigma meeting in Albuquerque. I won't be there, and I am sorry to miss your talk at the workshop. I spent the last week at the Chautauqua Institution, a delightful mixture of music, art, religion and science. Attached is a little piece that I wrote for the Chautauqua Daily. All good wishes to you and the students. As you may know, I got a splendid long letter with pictures from Leslie

Gilbert, one of the students who was in Costa Rica this year. Please give her my greetings if you happen to see her.

Yours ever, Freeman

Professor Dyson's Chautauqua article, "The Domestication of Biotechnology"[11] draws a parallel between the history of computers and the possible future of biotechnology. John von Neumann originally envisioned computers being large units similar to his original prototype, but within a few decades those large infrastructures evolved into the small personal computers that now form the heart of everyone's cell phone. Professor Dyson envisioned a similar development occurring with biotechnology. Starting in large institutional clinics, eventually it will land in the hands of hobbyists. Astonishing diversity will follow—envision a dog show with a class for modified breeds—bioengineered dogs with genetically purple hair, for instance. Such diversity drives technological ingenuity.

STS to FD:

22 November 2004

Dear Professor Dyson,

We trust that you and your family are doing well as 2004 winds down. For another semester, *Disturbing the Universe* has provided a springboard for fruitful discussions in our Science, Technology, and Society class….

1. Attached is a cartoon showing Ethics chasing the huge footsteps of Science, and Ethics shouting, "Wait Up!"[12] Is there a practical way to slow down the advances of science so that ethics can catch up?

FD to STS:

30 November 2004

Dear STS class,

Thanks to you all for the Thanksgiving message and the three questions that came with it. Nobody has found a better way to say thank you than the old Anglican prayer-book: "We thank Thee for our creation, preservation, and all the blessings of this life…." It does not say anything about roast turkey and stuffing. Here are some tentative answers to your questions:

1. Is there a practical way to slow down the advances of science so that ethics can catch up? My answer to this question is Yes. The picture shows science as a huge monster and ethics as a puny creature getting left behind. It is interesting that many people see it that way. Since I am a scientist, I see it differently. I see science as a collection of explorers trying to find their way through difficult country, with ethics as a distant beacon on the far horizon. There are many practical ways to slow down science. The easiest way is to stop providing the money that scientists need. The next easiest way is to impose rules and regulations that make scientists miserable. The third way is to hire lawyers and attack scientific enterprises with lawsuits. The fourth way is to attack laboratories physically and destroy experiments. All these four ways to slow down science have been used successfully. As I see it, the problem is to keep science moving ahead rather than to slow it down. Ethics should be a guide to keep it moving in the right direction, not a brake to slow it down. Of course science can be dangerous, but science guided by ethics offers much more hope than danger.

Another metaphor we considered for the relation between science and ethics envisions the large-stepping Science having an Ethics conscience riding on its shoulder, whispering advice and challenges in Science's ear such as, "*Should* we go there, just because we *can*?"

STS to FD:

2. We know that your family has top priority in how you spend your time. What advice would you give university students about to graduate, concerning priorities and expectations in balancing career with family life? How does this perspective at age 80 compare with one's perspective on these matters at age 25?

FD to STS:

2. I would not presume to advise any young person about priorities. As I learned from the Swiss nurse who helped deliver my oldest daughter, "Some people like to go to church and other people like cherries." Each of us has the freedom and the responsibility to choose our own priorities. My first piece of advice is, do not get trapped in your first choice of a job or a career. Always leave yourself room to change your mind and do something different. A good example is my fourth daughter, who became a Presbyterian minister. She always wanted to be a minister and was very good at it. She did a fine job as a solo minister for a little church in Maine. She took care of her parishioners and their problems, and her congregation grew. Meanwhile she also took care of her growing family. But after her fourth baby was born, she decided that she was stretched too thin. Her church and her family each needed more of her time than she had to give. So she had to make a choice. She gave up the church and is now a full-time mother. She does not regret the decision. She will be a full-time mother as long as she is needed, and then she can go back to being a minister at another church later. Of course she could not have had this freedom if she did not have a supportive husband. The most important advice I can give you is, be careful who you marry. I was lucky to find a wife who was a full-time mother for my kids. On the whole, I do not find that my priorities have changed much between the ages of 25 and 80.

STS to FD:

3. Another book we read this semester was Robert Pirsig's *Zen and the Art of Motorcycle Maintenance.* Pirsig describes teaching composition at Montana State University. One semester when he withheld grades, so that grades were no longer a superficial goal, the students' creativity was able to flourish.[13] Is there a similar principle that applies to the relationship between government regulations and developments in science or education?

Thank you for your time… We wish you and your family a Happy Thanksgiving!

The STS Class, Fall 2004

FD to STS:

3. Do government regulations have a stifling effect on science and education? I have no first-hand experience that I can bring to bear on this question. I think that the answer is generally no. The best way to answer the question is to look at different countries that have different ways of running science and education. For example, consider France and the USA. In France the government runs science and education with a tight control, in the USA both science and education are mostly run locally without detailed control by the federal government. The French system produces science and education that are on the average better than the USA. The USA system produces very uneven quality. The best American science and education is better than the French but the worst American science and education is far worse. So it is a question of values, whether you value fairness above freedom or freedom above fairness. The French system is more fair, the American system is more free. Of course, both systems put too much emphasis on grades, so we still need more teachers like Robert Pirsig.

That's all I have to say. Happy Christmas and New Year to you all… Yours ever, Freeman Dyson

13 The Varieties of Human Experience

The God whom science recognizes must be a God of universal laws exclusively, a God who does a wholesale, not a retail business. He cannot accommodate his processes to the convenience of individuals. —William James[1]

In January 2005 the Dyson family's New Year's letter included a breathtaking photograph of snow-capped Mount Shasta, taken from the snow-blanketed home of Professor Dyson's daughter Rebecca.

Fig. 13.1. Mt. Shasta, viewed from the home of Rebecca, Professor Dyson's daughter. Photo courtesy of the Dyson family.

This photo offered an excuse for another personal letter to Professor Dyson. I will not weary the reader with its details, but will offer only sufficient summaries and excerpts to provide context for Professor Dyson's response. As always, his replies offer glimpses of his character.

DN to FD:

> March 22, 2005
>
> Dear Professor Dyson,
>
> Thank you for your family's wonderful 2005 New Year letter…

> The breathtaking picture of Mount Shasta, taken from the home of Rebecca and Peter, motivates me to enclose a photo of Mt. Rainier taken last summer during our family vacation....

The bulk of the March 22 letter shared with Professor Dyson some reflections on his Witherspoon Lecture, "The Varieties of Human Experience," which he delivered on November 6, 2003, at the Center for Theological Inquiry in Princeton.[2] The title of Professor Dyson's speech echoes William James's Gifford Lectures on Natural Religion that James delivered at the University of Edinburgh. James's lectures were first published in 1902 as *The Varieties of Religious Experience.*[3] In his 2003 speech Professor Dyson described how "My lecture tonight will be squarely based on James's way of thinking." James's way of thinking makes a sharp distinction between institutional religion on one hand, and personal religion on the other hand. In his opening remarks Professor Dyson compared the approaches of William James and Sir John Templeton. He recalled how "James studied religion by studying the individual soul." In contrast, Templeton "believes that spiritual wisdom is to be found by combining the insights of religion with the tools of science." James focused on individual religious experience from a variety of personal perspectives, such as "The Religion of Healthy Mindedness" and "The Divided Self and the Process of its Unification." In his lecture Professor Dyson introduced two perspectives: "Theology and Theofiction" and "The Varieties of Neurological Impairment." As with all deep truths, no single perspective describes the entire picture. Professor Dyson generalizes the Principle of Complementarity that was introduced by Niels Bohr in the founding of Quantum Mechanics. "Complementarity means the existence of two pictures of a physical process that are both valid but cannot be seen simultaneously." In the quantum context, light passing through a diffraction grating behaves like a wave that spreads out to give an interference pattern; but when incident on a solar cell, light behaves like a rain of localized pellets. The question is *not*, "Which model of light is right?" Rather, the question is, "Under what circumstances does light behave like a wave, and under what circumstances does it behave like a particle?"[4] Professor Dyson goes on:

> *Following Bohr's broad use of the word, I propose that religion and science are also complementary. The formal frame of traditional theology, and the formal frame of traditional science, are both too narrow to comprehend the totality of human experience.*

Two windows. "Both frames exclude essential aspects of our existence. Theology excludes differential equations and science excludes the idea of the sacred." We need differential equations *and* a sense of the sacred.[5] "But the fact that these frames are too narrow does not imply that either can be expanded to include the other." Professor Dyson describes multiple panes within the two windows.

> *In fact science and religion belong to a wider array of human faculties, an array that also includes art, architecture, music, drama, law, medicine, history and literature. As an example of another perspective that leads to an increase of spiritual information, I can think of no better example than the work of Elaine Pagels…*

Pagels' research into ancient Christian manuscripts that did not make the cut for the scriptural canon "has given us a new picture of the Christian religion as it existed in early times before orthodoxies were rigidly imposed and heresies stamped out." Examples of Pagels' studies includes her treatment of *The Gospel of Thomas* as summarized in her book *Beyond Belief.*[6] Professor Dyson recommends

> *a program of support for scholars like Elaine Pagels who are learned in the languages and histories of other cultures and religions….All religions have a tendency to become rigid and intolerant…. If we could recover some of the ancient heretical literature of other religions…we might succeed in broadening the outlook of all religions.*

In my letter of March 22, about this passage I commented:

> In your lecture you said, "James looked at religion from the inside."…. Like Pagels, I find the church's meaning in its

sense of community.... For those personal relationships I am grateful. However, looking at a religion from the inside may distort the view almost as much as looking at it from the outside....

Elaine Pagels' book *Beyond Belief* became quite personal for me when I saw how it compares to *Gospel of John* to the *Gospel of Thomas*. For many years our denomination sponsored a Bible quizzing program at local, regional, and national levels.... I participated during my high school sophomore year.... That year we quizzed over the *Gospel of John*, King James Version. Reading Pagels' book replayed tapes in my head of familiar passages in John....

But Thomas has always been one of my heroes, because he dared to acknowledge his honest doubt.... Since quitting Bible quizzing and reading about the history of Western Civilization, one gathers that Christianity owes much to Platonic idealism and Constantine's conversion in addition to the Beatitudes of Jesus.... Ever since the great Schism between the Eastern and Western Church, and even more so after the Reformation, Christians have been splitting with boundless enthusiasm over points of doctrine; so it's not surprising to see it going on during the opening centuries of the Christian era.

You wrote that Pagels' glimpse of early Christianity "resonates well with a new generation of students who call themselves Christian but feel more at home with heresy than with orthodoxy..." Insofar as I have a theology, it would be like this: If a gracious Cosmic Mind truly exists, then to the person who honestly seeks to know it, that Mind will reveal itself in terms the individual can understand....

That's all for now about your Witherspoon speech. I share passages of it with students. "Science and Religion" seems to be an important topic to them... Most of the students are, like me, trying to figure out the religion into which they were born while observing it from the inside. In this struggle an insider... who appreciates his heritage but sups with a long spoon can make himself useful at a conservative Christian

> school. In fairness I must say that I find much tolerance among the administration and faculty here for insider heretics like me. I am not alone here....

In "The Varieties of Neurological Impairment Experience" section of Professor Dyson's Witherspoon Lecture, he describes how people who suffer from specific neurological disorders experience the world very differently from everyone else. In a sense, they are aliens among us. If their experiences were typical for everyone, the world would be organized to meet *those* needs, instead of the organization we accept as "normal." Theological understandings would also be different. Professor Dyson discussed autism as an example. Although autistic people have difficulty attaching meaning to stimuli such as facial expressions and body language, "in the autistic world, human beings love each other without understanding each other, and are incapable of hate. The theology of the autistic world must be radically different from Judeo-Christian theology." This section of Professor Dyson's speech on "aliens who may be living among us" again hit close to home.

The letter also included mundane questions that students wondered about.

> *Are you related to Sir Frank Dyson who, together with Arthur Eddington, organized the 1919 solar eclipse expeditions that confirmed Einstein's prediction of starlight deflection?...
>
> *I am frequently asked by students if you are related to James Dyson who markets the Dyson vacuum cleaner....
>
> Best wishes & warm regards, Dwight

Before he could answer the personal letter the Spring 2005 class sent Professor Dyson a few questions. He answered both letters at the same time. Here is the first part of his reply:

FD to DN:

7 May 2005

Dear Dwight,

Before I answer the students' questions, I take this opportunity to answer (better late than never) your splendid letter of March 22 with the picture of the family in front of Mount Rainier.... Your letter has a lot of meat in it. I won't try to answer all the thoughts and good ideas, only some of the questions.

I heard Elaine Pagels recently speak about the gospels of John and Thomas. One thing struck me forcefully. John is talking poetry and Thomas prose. Just from an artistic point of view, John is infinitely superior. Perhaps that is the reason why he is in the Bible and Thomas is not. The people who chose the canonical texts in the fourth century, like those who translated them into English twelve hundred years later, had a deep respect for style and language. Although I am closer in my view of Jesus to Thomas than to John, I have to admit that I would rather listen to John than to Thomas.

Answering your question about Dysons:

(1) I am not related to Sir Frank Dyson, but he came from the same part of Yorkshire as my father, and his father, who was a Baptist minister, married my grandparents. My father knew Sir Frank, and I heard a lot about his activities when I was a child. His fame certainly stimulated my interest in astronomy and science....

(2) I am also not related to James Dyson, but I once appeared side-by-side with him on a television program when we both happened to be in New Zealand. I liked him very much. One of my daughters has a Dyson vacuum cleaner and recommends it highly.

Asperger's is now the fashionable disease. It seems that half the people in Princeton either have Asperger's or claim to have it. I do not mean to belittle your son's problems, but he should not take the diagnosis too seriously. I would be interested to learn more about him.

In between our letter exchanges we also received a follow-up note from other correspondence:

FD to DN:

26 March 2005

Dear Dwight,

Thank you very much for sending the package with the splendid T-shirts. Thanks also to Krystal Smith for making them, and for her letter....

I just came back from a week in Phoenix where our daughter Esther has her computer conference and invites the family to join her. This time we had all our six children there and thirteen out of the fourteen grandchildren, ranging in age from ten months to fifteen years. So we had our Thanksgiving without any pumpkins and without any trick-or-treats. All good wishes to you and to Krystal. Yours ever, Freeman

Here are the questions that the Spring 2005 students asked:

STS to FD:

6 May 2005

Dear Professor Dyson,

Three dozen more students have completed our STS course, reading *Disturbing the Universe* as our first text and initiator of most class discussion....

1. In your seminar at Gustavus Adolphus University in 1999, your students arranged a debate over three possible policies for dispensing genetic technology. In your letter of 10 April 1999 you said you were "strongly in favor" of Option 3, "Allow genetic manipulation of embryos for any purpose, but only when the genetic resources exist to make it available to everybody." In exercising this option, would you make a policy distinction between somatic gene therapy and germline genetic engineering?

FD to STS:

26 March 2005

...Now for your students. First, thanks to them for the May 6 letter. Next, some brief answers.

1. I had not considered somatic gene therapy when I was discussing the regulation of genetic manipulation of embryos. I was talking about germline genetic engineering. That is what one is doing when one adds or subtracts genes from an embryo before it becomes a baby. Whatever changes you produce in the baby will be inherited by the baby's offspring. Somatic gene therapy is a different procedure altogether, done after a child is born, raising problems of safety and efficacy but not affecting the child's offspring. I would see no ethical objections to somatic gene therapy if the problems of safety and efficacy were solved. On the other hand, with genetic manipulation of embryos there are ethical problems even if the procedure is safe and effective.

STS to FD:

2. Do you see quantum computing having a major impact on modern life?

FD to STS:

2. Quantum computing is an exciting scientific problem. By exploring the possibilities of quantum computing, we are reaching a deeper understanding of quantum mechanics. Where that deeper understanding will lead is impossible to say. It is much too soon now to begin designing quantum computers or to predict what they may do. My own guess is that we will not have quantum computers standing alone, but quantum subroutines incorporated inside ordinary computers for doing special jobs. I do not see any major impact on the lives of people who are not computer experts. But I could well be wrong. It all depends on what kind of jobs the quantum processors may do. Your guess is as good as mine.

STS to FD:

3. This spring saw the passing of Hans Bethe. Besides being the physicist who taught everyone the nuclear reactions that make the stars shine, and a voice for arms control that "Administrations ignore at their peril,"[7] he was also your friend and mentor. Not only has the world lost a great mind that was coupled to a compassionate heart, but you have lost

a personal friend. What would you have us remember about Hans Bethe? (Our class "met" him in the documentary *The Day After Trinity*[8] in addition to your book.)

Thank you Professor Dyson for sharing your life, your insights, and—through your letters and books—your family with us as well. Thank you for explaining to we mostly "unscientific people the nature of the beast we are trying to control." [*DU* p. 5]...

-The Spring 2005 STS Class

FD to STS:

3. What would I like you to remember about Hans Bethe? The main thing was his delight in sharing whatever he was doing with anyone who was around. He loved solving scientific problems, whether they were important or unimportant. He loved exercising his skills as a calculator, but he kept his door open while he was calculating so that students and colleagues could come in and talk. At lunchtime he would always collect a bunch of students and go with us to the cafeteria. At lunch he talked a lot about the latest news in science and in the world outside. He was also a good listener and did not mind when somebody contradicted him and even occasionally proved him wrong. The thing I remember best from those days is that everyone called him Hans.

In conclusion I send you a poem that came this week from my eleven-year-old grandson Donald in California.

I am Donald Reid
I wonder if my dreams will come true
I hear the bubbling acid
I see the chemicals
I want to work on laser transportation
I am trying to be smart.

I pretend nothing
I feel the glass bottles
I touch the pencil that will sign contracts
I worry about my grades

I cry when I'm sad
I am trying to get into a good college.

I understand it will be hard
I say I can do it
I dream of being a famous professor
I try to do things I can't
I hope my dreams will come true
I am Donald Reid.

With greetings and good wishes to you all from the Dyson family.
P.S. In a separate message I send you a piece that my son George wrote about his friend Bob Hunter who died this week. Yours, Freeman

FD to STS:

7 May 2005
Dear Dwight and students,

Here is a piece written by my son George about Bob Hunter, the founder of Greenpeace. Greenpeace started as a group of friends of Bob in Vancouver, including my son and step-daughter Katarina. Katarina was the treasurer of Greenpeace when the total funds could be kept in a shoe-box. Now Greenpeace has two million members and is a major voice in world affairs. The first Greenpeace operation was an attempt to stop the testing of a five-megaton bomb on the Aleutian island of Amchitka. It is typical of my son that he had friends on both sides of the struggle.

I cannot resist including George's fascinating essay in Appendix 3. The range of interests and experiences of the Dyson clan is amazing. They may not realize it, but Professor Dyson's children have become co-teachers of STS.

STS to FD:

6 December 2005

Dear Professor Dyson,

We hope 2005 has been good to you and your family. During this fall semester of 2005, 44 more students have walked down the path of your experiences, and from your experiences have expanded their own. If at this busy time of year you can find a few moments to respond to the following questions, we would be delighted and appreciative....

1. If today you somehow fell into the Fountain of Youth and emerged as a 20-year-old, what would you pursue, and why?

FD to STS:

6 December 2005

Dear Dwight,

Thanks to you and the students for another thought-provoking lot of questions. 2005 has indeed been a good year for us and our tribe of fourteen grandchildren. We have much to be thankful for. Since time is short, I give you answers to your questions without much reflection.

1. I just spent four exciting days lecturing at St. John's College in Annapolis, a remarkable place (founded in 1696) where the students study a purely classical curriculum based on the Great Books.[9] *Those students asked me the same question. I told them I would still have the same problem that I had sixty years ago when I was a real twenty-year-old. My problem is that I have many interests, biology, medicine, astronomy, physics, literature, history, languages, but only one talent, mathematics. So I would do the same thing that I did sixty years ago, looking around for ways to use my talent to explore interesting problems in many different fields. Sixty years ago, the most interesting problems were in physics, and so physics was my first playground. Today the most interesting problems are in biology and I would probably concentrate my efforts on those, trying to understand the deep structure of cells and brains and genomes and their relationships with one another, using my mathematics to dig a little deeper into the mysteries.*

STS to FD:

2. Your friendship with Edward Teller is all the more remarkable because you and he became firm friends in 1955, about a year after the infamous Oppenheimer security risk hearings. We know that, at that time and thereafter, many physicists refused to shake Teller's hand. Yet you were able to rise above the situation, and become Teller's friend while remaining Oppy's friend also. [*DU* p. 93] And while on the Princeton Citizen's Committee, you and Emma Epps[10] became good friends even though you disagreed with her on your recommendations as Committee members. [*DU* p. 179] How were egos disarmed when you became a good friend with people with whom you disagreed?

FD to STS:

2. I never had any difficulty in making friends with people that I disagreed with. Life would be very dull if we could only have friends who agreed with us about everything. Actually I disagreed more strongly with Emma Epps than I did with Edward Teller, but that made no difference to our friendships. I liked and respected both of them equally. Emma was hostile to science because she identified science with Princeton University and Princeton University had been treating black people badly for two hundred years.[11] *She just did not trust scientists to use their power wisely. That was understandable and maybe she was right. Whether she was right or wrong, her presence as a spokesman for the opposition made our meetings much more meaningful. The whole point of the meetings of our citizens' committee was to give all sides of the debate a chance to be heard. Emma's presence made it clear that the opposition was not only heard but also treated with respect.*

With Teller my disagreements were more superficial, about details of the work we were doing together on nuclear reactors. I did not disagree with him about his testimony at the Oppenheimer security hearing. What he said at the hearing was an honest statement of his opinion. I was at that time still British and not involved in American security problems. I thought the main issue in the Oppenheimer hearing was whether the same rules should apply to the famous people at the top as to the little people at the bottom. If anyone who was not famous had behaved as Oppenheimer behaved, telling lies to security officers and making up stories to confuse them, he would certainly have been refused clearance. So to me the question was, should the rules be applied fairly to big shots and little shots alike? If the rules were applied fairly, Oppenheimer certainly should lose his clearance. It seemed to me reasonable for Teller to say what he thought about this. Anyhow, Teller was a likeable character and remained a close friend of Leo Szilard who disagreed with him much more fiercely than I did.

STS to FD:

3. What was the greatest intellectual community with which you have been connected?

FD to STS:

3. The greatest intellectual community I have known was the physics department at Cornell University when I came there as a student in 1947. It was full of brilliant people, starting with Hans Bethe and Richard Feynman, and it was a real community with old and young people helping each other out. It was a small enough group so that we all knew each other personally and felt like a big family. Also we had theorists and experimenters and engineers all working together. The isolated situation and harsh winter climate helped to bring out people's best qualities. You could not survive without help, and so everybody helped.

STS to FD:

4. At what time in your life did your greatest intellectual growth occur?

FD to STS:

4. My greatest intellectual growth was at the same time that I described in question 3, when I was a graduate student at Cornell aged 24 to 25. That was when I learned from Bethe and Feynman how to solve serious problems in physics, and I was also for the first time outside England and making friends with people from many countries.

STS to FD:

5. "...A battered old Dodge convertible with the roof open...careening at breakneck speed down through the institute woods to the river..." [*DU* p. 75] What is your fondest memory of the immaturities that are the privilege of youth?

... We wish you and your family a Happy Christmas season and a joyous New Year.

Warmest regards, STS class, Fall 2005

FD to STS:

5. It is impossible to say which is the fondest among many memories of immaturity. Perhaps the fondest of all is night-climbing over the ancient buildings of my high-school in England, when war-time blackout was complete and the only light was moonlight. I used to go climbing the crumbling stone towers of our buildings with my friend Peter in the early hours after midnight. We did not bother with such frills as helmets or climbing ropes. It was wartime and we took crazy risks, hanging onto medieval stone saints over a hundred-foot drop. Peter was killed two years later as a parachutist in the battle of Arnhem. I must be one of the few surviving people who remember him.[12]

That's all for this time. Thanks again for your interest and your good wishes. Happy Christmas and New Year to all of you. Yours ever, Freeman

14 Talk to Your Enemies

We must learn to listen to one another, and to understand one another.
—Mikhail Gorbachev[1]

STS to FD:

12 October 2006

Dear Professor Dyson,

I trust you and your family and all those wonderful grandchildren are doing well. I must apologize for the failure of my spring 2006 semester of Science, Technology, and Society to send you any questions....

They were actively engaged on other ways however. You may enjoy a few responses from students. On the final exam I say, "Describe the topic covered in this class that has meant the most to you personally. Full credit will be given for your genuine, thoughtful response." A young man wrote,

"So much could be said, yet I find a great home in Dyson's Principle of Maximum Diversity, that the universe is designed to be as interesting as possible. So often do I find others, as well as myself, trying to imitate those we hold to be of a certain integrity or quality, yet find little room to be ourselves in a society driven by pop idols and classical stereotypes…"

This student refers to Professor Dyson's 1986 speech to the Conference of Catholic Bishops:[2]

Why is the world so unjust? What is the purpose of pain and tragedy? I would like to have answers to these questions, answers which are valid at our childish level of understanding even if they do not penetrate very far into the mind of God. My answers are based on a hypothesis which is an extension both of the anthropic

> *principle and of the argument from design. The hypothesis is that the universe is constructed according to a principle of maximum diversity....It says that the laws of nature and the initial conditions are such as to make the universe as interesting as possible.*

Another student wrote,

> "Whatever technology we choose, we are stuck with it—the Magic City [*DU* Ch. 1]....Also, technology seems to now be economic-driven rather than help-driven. Take cell phones. About every month a bigger, better model comes out. Would it not be more beneficial to produce a new one only once a year? This only feeds consumerism..."

This comment reminds us of the values of the Mennonites. Even though many of them still depend on horse-and-buggy transportation, they are not dogmatically opposed to technology in general. When offered a new technology, the question they ask themselves is *not* "Is this more efficient or convenient?" Rather, they ask a more important question: "Will this contribute to community?" It seems that the larger society makes convenience the highest priority. The question "Does the First Commandment[3] apply to the God of Convenience?" arises many times each semester.

A third student wrote

> "After attending classes and trying to answer questions about life, I was somewhat embarrassed to not have thought about them before. I've also come to realize that I like to ask questions, but am usually not prepared to seek the answer. What I mean by this is that I'll ask a question and if I can't immediately find the answer, then I'll drop the question. While trying to answer several questions in this class I realized that many times the answer isn't as important as the journey to get there."

This student reminds us of your comment on working with Hans Bethe: "I was now ready to start thinking." [*DU* p. 49]

SNU is still taking students to the Quetzal Education and Research Center in the Talamanca Mountains of Costa Rica.... Those who are supported by a Dyson Scholarship are supposed to send you a thank-you message and tell you about their work....

Stay well, and best wishes to you, Imme, and all your extended family.

Warm regards,

Dwight

PS- ... Does George still have his tree house in the Douglas fir?

FD to STS:

13 October 2006

Dear Dwight,

Thank you very much for your long message, full of ideas and questions. I do not have time to answer it adequately. I will go through the questions quickly. I enjoyed the students' comments...

Please don't ask the students who go to Costa Rica to write thank you letters. I am happy to get such letters, but only if they are spontaneous.

George's tree-house still exists after thirty years. He built it well and solidly. But he has not used it for many years and he has made it inaccessible so that it will not attract children. Since it is a hundred feet up, it could be a cause of tragic accidents.

Imme is now on her way to Katmandu for two weeks of trekking with one of her running friends. This is a celebration of her seventieth birthday. Luckily she found a companion, as I am too old to go with her. Just now, if all went well, they are staying overnight in Bangkok, and will reach Katmandu tomorrow.

Thanks and greetings to you and the students. Yours ever, Freeman

On 17 October 2006, in view of nuclear weapons testing by North Korea, we wrote to Professor Dyson earlier than usual. Given his study of nuclear weapons policy through his service at the Arms Control and

Disarmament Agency, [*DU* Chs. 12–14] his book *Weapons and Hope*,[4] and his work with the Jasons, he could offer informed perspective regarding North Korea's brazen development of nuclear weapons. His reply offered an unexpected insight:

FD to DN:

17 October 2006

Dear Dwight,

Your message arrives just as I am leaving for California, so I answer it briefly. Here is what I would say to the students.

Don't believe the media hype about the North Korean bombs. Whether they have bombs or not is not important to the United States. Nuclear bombs are important to the North Koreans as a political status symbol. We only increase their value to the North Koreans by making such a fuss about them. The best thing for us to do is to ignore them. If in the future we want to get rid of North Korean bombs, it can only be done by international agreement, and then we must also be prepared to get rid of our own.

Yours sincerely, Freeman

Given the huge ego of North's Korea's supreme leader—an affliction that seems to strike all dictators and wanna-be dictators—this recommendation makes sense. Kim Jung Il is a tyrant by inheritance who evidently cares nothing about the North Korean population. Like tyrants throughout history, it's all about *him*, so he must be delighted to see people in the States and elsewhere getting excited whenever he launches a rocket that could carry nuclear weapons. He knows how to yank our chains. Professor Dyson's reply about North Korea reminded us of his comment in *DU* about the Soviet anti-ballistic missile systems:

> *My study of the Soviet literature convinced me that the Russians were totally serious about maintaining the superiority in conventional forces, infantry, tanks and guns that had brought them victory in World War II… They were not, in the same sense, serious about the advanced technology that dominated American thinking… Khrushchev's ABM system was only the latest example of the Soviet tradition of defense by bluff… The Soviet leaders were*

> *able, without actually lying, to exaggerate their strength and distract attention from their weaknesses. … Even if I had known in the summer of 1962 what was to happen in October,*[5] *I could never have hoped to persuade the senior officials in ACDA to accept my opinion that the missiles in Cuba were only a typical defense by bluff, which Kennedy was under no obligation to demolish. When we demolished the Soviet missile bluff as conspicuously as possible with public statements of the results of U-2 photography, we forced the Soviet Union to replace its fictitious missile force by a real one.* [DU pp. 136–137]

Perhaps those lessons from history should influence our present challenges regarding nuclear weapons. On the other hand, in the *DU* passage just cited Professor Dyson went on to describe how a policy of ignoring the bluff would not survive American politics. No administration could allow itself to be perceived as ignoring an adversary's apparent nuclear buildup, even if that buildup was mere bluffing. In politics, perception often outweighs reality.

STS to FD:

> 4 December 2006
>
> Dear Professor Dyson,
>
> We hope that you and your extended family are doing well this fall. May the holiday season be a time when you will be able to hold many grandchildren. If you have a few moments to respond to a couple of questions from yet another Science, Technology, & Society class, we as readers of *Disturbing the Universe* would appreciate it so much.
>
> In the May/June 2003 issue of the *Bulletin of the Atomic Scientists*, we found an article by Peter Hayes and Nina Tannenwald called "Nixing Nukes in Vietnam."[6] The article recalls a scene that we read about in *Disturbing the Universe*, where some Pentagon official remarked that "I think it might be a good idea to throw in a nuke from time to time, just to keep the other side guessing." [*DU* p. 149] The *Bulletin* article describes how four members of JASONs: you, Robert Gomer, Courtenay Wright, and Steven Weinberg, were so appalled

by the remark that you decided the best way to counter it would be to conduct a serious study of the effects of "tossing in a nuke." Here are our questions:

1. Given the present Administration's demonstrated willingness to use preemptive strikes against other states, as a co-author of the 1966 JASON study mentioned above, what advice would you give the Administration—and we citizens—in today's global political environment?

FD to STS:

9 December 2006

Dear Dwight and students,

I just came back from a tour of family in various places, our minister daughter Mia in Maine, our son George in Bellingham and our step-daughter Katarina in Vancouver, with their various offspring. A great trip. Came back yesterday to find your message.... Here are answers to the questions.

1. I find it encouraging that even an administration as immoral and incompetent as our present one has not been talking about "tossing in a nuke" as Maxwell Taylor did in 1967. Even Mr. Bush now understands that tossing in a nuke would not help to solve his problems in Iraq or in Iran or in North Korea. The advice I would give to the administration today is: talk to your enemies. There is one person still around today who knows how to run Iraq, namely Saddam Hussein. Take him out of jail and make use of him, as we did with the Japanese emperor in 1945. Give him the respect that is due to him as a head of state, and let him try to hold the country together. If he fails, if Iraq falls apart into three countries, with the Shia and the Kurds seceding, that is his problem and not ours.

Alas, as we know with sorrowful hindsight, the George W. Bush Administration did not see fit to talk with its enemies, but chose instead to embark on a nation-building project after starting a war on false pretenses. Back in January 1961, President Dwight Eisenhower used his farewell speech to warn Americans against buying into the "military-industrial complex."[7] The United States has been at war with somebody almost continuously ever since the end of World War II. War is a gargantuan, profitable business.

The Jasons are freelance consultants whose typical clients include the Department of Defense. Given the war that had just been launched in Iraq, we thought Professor Dyson, as a Jason, might see in that war a possible Jason study, thereby motivating the next question.

STS to FD:

2. What topics should be top priorities for a JASON study at present?

FD to STS:

2. Topics for Jason studies always depend on having a sponsor who pays for the study and wants to listen to our advice. If nobody in the government will pay for it, then nobody will listen to it. A Jason study of military strategy in Iraq would not be useful. Even though that is the most important subject, we do not have the credentials to give advice about it. The main topics for our upcoming Winter Study in January are the RRW (Reliable Replacement Warhead), and the protection of buildings against cars and trucks loaded with explosives. These are both technical problems for which we have the credentials to be taken seriously. As you can imagine, the Oklahoma bombing[8] provides a lot of the data for the building study. The problem is to make a building safe without making it look and feel like a prison.

The RRW is partly a technical problem and partly political. The technical problem is to rebuild our nuclear weapons with control systems built into the structure so that it is physically impossible for anyone who steals a weapon to explode it. If this were done to the weapons in the stockpiles here and in Russia, it would make the world a bit safer. It would mean that nuclear terrorists would have to build their own weapons. The political problem with the RRW is that it looks as if we are developing new nuclear weapons, and so gives an excuse for other countries to develop new weapons too. On the whole I am against the RRW for political reasons, but still it makes technical sense and I agreed to work on it. The good thing about Jason is that we are each free to choose which problems to work on.

Professor Dyson and Steven Weinberg were both members of the Jasons. Among his many accomplishments, Professor Dyson is known throughout the physics community for his proof of the equivalence of the Feynman and the Schwinger-Tomonaga ways of doing quantum electrodynamics. Among his many accomplishments, Professor Weinberg is known throughout the physics community for developing the electroweak interaction model that unifies quantum electrodynamics and the weak interaction. Freeman Dyson and Steven Weinberg were good friends.

STS to FD:

> 3. Did you and Steven Weinberg ever discuss between yourselves the topics that you address in the final chapter, "Dreams of Earth and Sky" [*DU* Ch. 24] and in your review article, "Physics and Biology in an Open Universe?"[9] Weinberg's position on questions that belong to religion and not to science are described in Chapter 11 ("What About God?") of his book *Dreams of a Final Theory*.[10] There he beautifully articulates the issues involved, with a clarity we seldom hear from the pulpit. We are wondering if you and he ever had the opportunity to discuss these matters between yourselves, in addition to the discussions that occurred in public. (It's inspiring to see that you and Steven Weinberg have dreams and are willing to share them.)
>
> … We know you are very busy and we apologize for any subtractions our message takes from your time with your friends and family. We wish you a joyous holiday season.
>
> Thank you again, and best wishes, STS class, Fall 2006

FD to STS:

> *3. I am sorry to say that Steven Weinberg and I never sat down to discuss our disagreements privately. It seems we are both more comfortable arguing in public than in private. We are and remain good friends, and don't want to let our disagreements damage our*

> *friendship. I think everything we have said about philosophical and religious questions is out in the open where you can read it for yourselves.*
>
> *That is all I have to say about the questions. Thanks again for your thoughtful remarks. I wish all of you a joyful holiday and safe travel. Yours ever, Freeman*

An unexpected letter dated 21 March 2007 arrived from Hyung Choi, a theoretical physicist at St. Edmunds College in Cambridge, UK. Professor Choi was working on program development for the Templeton Foundation. He was scheduled to meet with Professor Dyson at the Institute for Advanced Study on March 26. The Foundation was considering the establishment of a prize named in Professor Dyson's honor "that could be given to a number of young scholars who are working on the fields of his life-time interests." Dr. Choi wrote, "Knowing that he is a modest person who could say 'no' to an idea that may bear his name," our opinion was solicited on whether such a proposal to Professor Dyson would be a good idea.[11] Instructed to keep the matter confidential, we answered Dr. Choi immediately in the affirmative, noting that "Professor Dyson is a modest person. He would not endorse any program that is about stroking someone's ego. But he may be interested in helping young scholars carry forward the kind of work in which he invested his life." On the 27th Dr. Choi reported that he had a "very good meeting with Professor Dyson," who was "quite positive."

15 God Has a Sense of Humor

"There goes a toucan!—a flying banana. Seeing that makes you think God has a sense of humor." —Leo Finkenbinder[1]

Near the middle of the Fall 2007 semester we sent Professor Dyson more samples of the STS weekly letters that form a portion of student assignments. The students are to imagine the STS course to be a journey, and reflect over it by writing to someone at home.

FD to STS:

13 October 2007

Dear Dwight,

Today I finally had time to read the student letters that you sent. Thank you very much for sending them. Many of them show remarkable insight, besides a good command of the art of writing. You are doing a splendid job getting these kids to reflect on who they are and what they may become. I will not comment on them individually.... I will be happy if you keep on sharing them when they are responding to Pirsig instead of to me. I look forward to seeing the same writers grow and change as time goes on. Thanks again, yours ever, Freeman

As the Fall 2007 semester drew towards its end, it was time for us to offer a few questions to Professor Dyson.

STS to FD:

November 12, 2007

Dear Professor Dyson,

We hope that the 2007 Thanksgiving season finds you and all your family doing well... We thank you for the contribution you made to help students get to SNU's field station in Costa Rica... Our Costa Rica field station has recently become a central site for quetzal studies through an agreement between SNU and Arizona State University[2]...

If we may, we'd like to present you with a few questions.... Thank you for your consideration.

1. You speak of art, opera, music, and literature in parallel with science in your book. We were curious as to where your love for the arts began in your life, and who are some of your favorite artists, musicians, and authors?

FD to STS:

23 November 2007

Dear Dwight,

Thank you for two items. First the package of papers which I will not comment on now. I enjoyed very much reading them. Second your letter of November 12 with questions from your STS class. Here are my responses to these. As usual, my responses are brief and not profound. Each of the questions would take several pages to answer adequately. These responses are just comments and not answers.

1. I was exposed to a great deal of music as a child, as my father was a professional conductor and composer. I was dragged to many concerts. I was more interested in the musicians than the music. When I was about five years old, one of the ladies at a concert said to me, "Aren't you lucky to get to hear so much music?" and I replied, "Music is very nice but too long." She told that to my father and he was highly amused. After that he always supplied me with a score so I could read along instead of listening to the music. Music has always been a foreign language to me, interesting to observe but basically incomprehensible. The same is true of art and opera.

Literature is different. Literature speaks to me directly. I acquired a taste for literature from my mother, who was even more tone-deaf than me. As a child I disliked boy-stories and loved girl-stories. My favorite books were The Wizard of Oz, Alice Through the Looking-Glass, Little Women, The Secret Garden, all with girls for heroes. The Magic City was an exception, but the real hero of that story is the nurse-maid and not the boy. Later when I had children of my own I loved to read aloud Charlotte's Web. I was always more at home with literature than with art or music. I enjoyed writing more than drawing or playing the violin.[3] *In the end I slipped easily from doing science to writing books.*

A recurring discussion theme in STS observes that any technology can be an instrument of good or it can be leveraged for evil. Does any technology exist that has not been weaponized?

STS to FD:

2. Do the good deeds that technology makes possible vastly outweigh the evils? Should we have not pursued technological advancement as a society at all, and maintained a simpler way of life, free of the possibility of nuclear war or genetic monsters?

FD to STS;

2. The choice whether to adopt a technological lifestyle was made by our ancestors ten thousand years ago when they invented agriculture. Agriculture was technological from the start, and led directly to civilization, with all its advantages and disadvantages. Yes, I think the advantages greatly outweigh the disadvantages. Before agriculture, we were hunter-gatherers living in small bands and incessantly fighting our neighbors. Paleontological evidence indicates that we were dying violent deaths then, much more frequently than we are today. There was never a time when we were living a simpler way of life without violent quarrels and slaughter. To get rid of nuclear war and genetic monsters will not be easy, but it will probably be easier in a civilized world than in an uncivilized world. The idea that we could go back to a peaceful world by getting rid of technology is an illusion.

After the September 11, 2001 terrorist attacks on the World Trade Center in New York City and the Pentagon in Washington DC, the Administration pushed for and the Congress passed legislation called the "Patriot Act."[4] In class discussion we observed that "patriotism" can be a loaded word. Seeing beyond mindless flag-waving, an authentic patriot calls out a nation's failings and injustices in order to help the nation improve. To celebrate a nation's achievements and virtues while ignoring its failures and atrocities is not patriotism, but propaganda. As one author put it, the so-called "War on Terror" after 9/11 "turned into a war on American ideals."[5]

STS to FD:

> 3. What are your feelings about the Patriot Act? Do you believe it infringes on basic human rights? We ask this question in light of possible parallels with the McCarthy hysteria of the early 1950s.

FD to STS:

> *3. Yes, to me the Patriot Act is a big step backward. It is just as Benjamin Franklin said, people who give up a little freedom to obtain a little safety end up by losing both freedom and safety. Our present-day Patriot Act and our bad treatment of prisoners are even worse than the hysteria of the McCarthy era. McCarthy tried to put innocent people in jail, but he did not try to hold them in jail without trial, and he did not advocate using torture to make them talk. I think this country is behaving much worse today than we were in the 1950s, and the rest of the world knows it. Our bad treatment of prisoners dismays our friends and encourages our enemies.*

The next question was motivated by our class discussions of nuclear weapons and the Cold War. Those discussions close with a look at attempts by the Air Force to place offensive weapons in low-Earth orbit. Ever since the first satellites were launched in the late 1950s, they have been used for military surveillance and communication. When extolling the virtues of weapons-laden platforms in orbit, Air Force General Lord wrote, "It's the American way of fighting."[6] An article by Nina Tannenwald[7] points to the inevitability of a new arms race comparable in costs and risks to the Cold War, as soon as any one nation places offensive weapons in orbit. Other than a few such articles, very little public discussion has been forthcoming about the prospect of the USA taking the initiative in placing weapons-laden satellites in low-Earth orbit. This echoes the history of the hydrogen bomb, where the decision to build and deploy thermonuclear weapons was made by a small circle of officials with little public discussion.[8]

STS to FD:

4. Can you see any possible reason for the USA to attempt to put weapons in space?

FD to STS:

4. I don't know what "weapons in space" means. I would say we already have huge numbers of weapons in space. At the moment we have hundreds of military satellites in space, giving our army and navy the ability to communicate and navigate and aim missiles accurately. Without these satellites, none of our army and navy units could function. Do these satellites count as weapons? I don't see why not. The big problems will come when our enemies begin shooting them down. It will be much easier to shoot them down than to put them up. In the long run I think we have only two options. Either negotiate an international treaty allowing such weapons to everyone with the right of free passage, or stop relying on them and gradually withdraw them.

The Human Genome Project, begun in 1990 and completed in 2003, was on the minds of the Fall 2007 students.

STS to FD:

5. With the advent of genetic engineering, should we attempt in any way to alter the human genome, especially if that change would be passed down to the patient's children?

FD to STS:

5. There are two main reasons for parents to want to alter the genetic endowment of their babies. The first reason is to get rid of genes that are known to cause fatal or incapacitating diseases such as Tay-Sachs or Huntington's or cystic fibrosis. The second reason is to give the baby a better chance to compete in life by having genes for superior intelligence or superior ability to play football. The first reason is a cheap and effective treatment for otherwise incurable diseases. The second reason is a cheap and effective entrance ticket to high-income professions. The two reasons raise very different ethical problems. Many people would like to allow the first and forbid the second. I would prefer to allow the parents freedom to make genetic changes for both reasons, if and only if the procedure is medically safe and is equally available to rich and poor. I don't expect all of you to agree with this.

STS to FD:

6. When discussing ground-breaking genetic technologies, what do you think it means when people assert that we should not pursue a certain course of action any further because we would be "playing God"?

Thank you for the contributions you have already made to our class. We wish you and your family a wonderful Thanksgiving and Christmas season.

Science, Technology, and Society class, Fall 2007

FD to STS:

6. This is the same question as number 5. Every medical doctor who advises a patient making a life-or-death decision is playing God. So is a minister who advises a parishioner. I have two daughters who are medical doctors and one who is a Presbyterian minister, and they all do this from time to time. In my opinion, the ethical problems are the same, whether the doctor or the minister is using old-fashioned professional judgment or new-fashioned high technology. The patient or the parishioner comes to you confused and scared, and you try to help the patient or the parishioner to look at the problem objectively and calmly. I would say that is playing God, and the best doctors and ministers are those who can do it well. It is useful to remember that God also has a sense of humor.

That's all for today. Please say thank you to the students for signing your letter. That gives me a glimpse of who they are. I find it interesting that seven of them are theology majors and seven are science majors, with the big majority doing more practical things.

Since Thanksgiving is over, I wish you all Happy Christmas. Yours ever, Freeman Dyson

STS to FD:

26 November 2007

Dear Professor Dyson,

Thank you once again for taking the time to answer our questions.... Best wishes for a wonderful Christmas season that is almost upon us....

FD to STS:

19 December 2007

Dear Dwight,

I suppose the students have gone home, but I ask you anyhow to thank them for their two Christmas cards. And thank you to yourself for the latest batch of essays. The latest batch was particularly illuminating. Since these were the last essays of the course, the students were writing more from the heart than they were before. And I found their struggles with fundamentalism very moving. I am not supposed to grade them, but I have to tell you that I liked especially Sarah Bean. I was happy to see that she is intending to become a minister like my daughter Mia. Since she has finished the course, you might tell her privately that I read her essay and liked it very much.

Happy Christmas to you and the students. Yours ever, Freeman

16 Family First, Friends Second, Work Third

Every man's faithfulness to his band was strong, but of all human ties that of family was the strongest.

—Luther Standing Bear[1]

The Dyson New Year Letter of 2009 related an amusing story of the adventures of Freeman and Imme on their visit to the Galápagos Islands.

Fig. 16.1. Freeman and Imme Dyson, with friends on the Galápagos Islands. Photo courtesy of the Dyson family.

The airline lost Freeman's suitcase, but on the ship Freeman and Imme made friends with four young women who took it upon themselves to look after this spunky elderly couple from Princeton. Freeman and Imme teasingly called these delightful young women the "Gang of Four." Since Freeman had no extra sets of clothes, the Gang of Four lent Freeman some of their blouses to wear, and they "fitted him perfectly." Many people would think of losing a suitcase as a vacation-spoiling disaster. Professor Dyson wrote that losing a suitcase is a good way to make friends.

STS to FD:

13 April 2009

Dear Professor Dyson,

I bring you greetings from the Spring 2009 section of the Science, Technology, and Society class.... Thank you for the wonderful 2009 New Year Letter from you and Imme. I shared the letter with the students. The trip to the Galápagos Islands sounds so splendid... One can imagine a sense of presence while being on those islands among its creatures while hiking in Darwin's footsteps. It sounds like your "Gang of Four" took good care of you during the trip too! I can imagine your and Imme's spunkiness drawing them to you....

Your New Year letter mentions your grandchildren. My oldest son and his wife made us first-time grandparents last September. It's so much fun to see our six-month-old grandson discovering the world. We hope he always laughs as spontaneously as he does now.

Back to the STS class. We have read *Disturbing the Universe* and the students have a few questions they would like to ask you if that would not be too excessive an imposition...

1. Is there anything for which science does not have the answer?

FD to STS:

17 April 2009

Dear Dwight,

Thanks to you and the students for your friendly response to my little newsletter. The next newsletter will be full of another adventure, a family reunion at the Russian space-launch center in Baikonur, Kazakhstan. My wife and I came back from this trip three weeks ago and are still absorbing the impact of it. The Russian space culture is radically different from ours. For us space-travel is an adventure. For them it is a vocation.... Now let me try to answer the students' questions.

1. Here I can answer simply "almost everything." Science is a bag of tools which are spectacularly effective in answering questions for which the tools are designed, but quite ineffective for answering questions outside that area. Examples of questions outside the scope of science are: Which is the best form of government? Which is the best way to distribute wealth among humans? Do animals have rights equal to human rights? Is there such a thing as a just war? Is Sharia, the Islamic system of law, acceptable as the basis for Islamic civic societies? And so on, and so on. All questions about values rather than quantities.

STS to FD:

2. If you could select one individual from history for a face-to-face conversation, who would you select and why, and what might you ask them?

FD to STS:

2. The obvious choice would be Jesus of Nazareth, the person who had the most profound effect on our history and about whom we know very little. I would like to listen to him talk rather than asking him questions. If you insist on a question, I would ask whether he would consider it good or bad for the religion that he founded to become the official religion of the Roman Empire.

STS to FD:

3. What should the generation now young be most concerned about?

FD to STS:

3. The generation now young should have many different concerns, depending on their individual circumstances and tastes. One size does not fit all. One of the concerns that should receive much more attention is the abolition of nuclear weapons. I think the time is ripe for the generation now young to take this problem seriously and do something about it.

STS to FD:

4. For what would you like your grandchildren to remember you?

FD to STS:

4. I would like my grandchildren to remember me as I am, a friendly old codger who enjoys watching his grandchildren grow.

STS to FD:

5. In your New Year's letter, you and Imme wrote of the events of last January 20 [the inauguration of President Barack Obama, the first Black president of the United States], "The scene in Washington reminded Freeman poignantly of a day long ago in Washington when he was marching with another crowd in the same place." We wonder what went through your mind last January 20, since you were at the opposite end of the Mall with the marchers that day in 1963 when Martin Luther King delivered his "I Have A Dream" speech.[2] We would love to be offered a glimpse into your memories, with whatever you might want to say in addition to what you already expressed so eloquently about that day in *Disturbing the Universe*. [*DU* pp. 140–141]

Thank you again Professor Dyson for all you have contributed to our lives. Best wishes to you, to Imme, and to your beautiful family.

-Spring 2009 STS class

FD to STS:

5. About the Washington March of 1963,… I was inspired by the marvelous fact that in this young prophet Martin Luther King the force of black liberation was combined with the gospel of non-violence. This was for me the dream that made sense. Later, after King was assassinated, the dream was for a time lost. But now the election of Obama seems to have brought it back to life.

I am sorry these answers are brief and inadequate. I would have more to say if I could meet with you face to face.

Congratulations to Dwight for his new grandson. I wish him as much joy as we have with ours. With best wishes for the future, for you and for him. Yours ever, Freeman

DN to FD:

17 April 2009

Dear Professor Dyson,

Thank you for kindly responding once again to our questions.... The textbook author ceases being some mysterious "they" and becomes "our friend, Professor Dyson."...

Students from SNU are still going to SNU's Quetzal Education & Research Center in Costa Rica.... Your generosity is at work here.

A couple of weeks ago we watched the NOVA documentary "Richard Feynman: Last Journey of a Genius"[3] which tells about his and Ralph Leighton's efforts to go to Tuva.[4] Now you and Imme are going to Russia. Such a spirit of adventure keeps you young at heart. I repeatedly encourage these students to travel and see the world at every opportunity, such as immediately after graduation. "When else," I ask, "are you going to be this relatively free to go?"...

What is George doing these days?.... Is he still building boats...? I hope he is still writing books. Writing books seems to be something the Dyson clan does very well.

.... Thanks for the compliment about our new grandson .... You are in a position to know about the joy of grandchildren, which makes your good wishes for us all that more valuable. Your wonderful grandchildren make you and Imme wealthy in the ways that really matter.

I'll send you soon some of the students' weekly essays...

Best wishes, Dwight

FD to STS:

17 April 2009

Dear Dwight, thanks for your news about yourself and the students. Answer to your question about George. He still runs his

boat business but his main activity is writing a biography of John von Neumann.[5] *As always, he is doing a careful job as a historian and spends a lot of time in Hungary. Recently we had a ceremony at Princeton to install a bronze portrait of von Neumann at the building where he built his computer. The Hungarian ambassador came to the ceremony, and I was introduced to the ambassador with the words, "This is George's father." So I glowed with pride. The last time we saw George was at the family reunion in Kazakhstan.*
Yours, Freeman

STS to FD:

29 April 2011

Dear Professor Dyson,

We trust that 2011 so far has been a splendid year for you and your extended family. It has been far too long since one of our Science, Technology, and Society classes has ventured to ask you some questions.... Thank you for your consideration.

1. You wrote that, in the long term, qualitative decisions are more important than quantitative ones. [*DU* p. 192] What are some of the important qualitative decisions being made today?

FD to STS:

29 April 2011

Dear Dwight,

Sorry, the timing could not be worse. Your thoughtful letter and the students' questions arrive just as we are leaving for a week in England, and your semester ends the day after we get back. That is not your fault, just bad luck. ...I will give quick and superficial answers to the questions. I am sorry I cannot do better. Humble apologies to the students.

1. I suppose the most important qualitative decision being made today is whether equality of opportunity applies to communities or to individuals. In the USA recent legal decisions have outlawed affirmative action programs giving preferential treatment to communities. The result is increased inequality at all levels of society. I consider this an unmitigated evil.

STS to FD:

2. Do you think that scientists are becoming more—or less—sensitive to ethical issues regarding their work than when you wrote *Disturbing the Universe* in 1979?

FD to STS:

2. I do not see any big change in ethical sensitivity of scientists. The great majority of scientists have nothing to worry about because their work has no serious effect on human affairs. Those who work on nuclear weapons or medical trials have generally been sensitive to the ethical consequences. I do not know whether their sensitivity has increased since 1979.

The next question was motivated by contemporary events in the Middle East and North Africa. Those rebellions against authoritarianism, driven especially by young people in those nations, came to be known as the "Arab Spring" of 2011. Among the results at the time, the long-term Libyan dictator Muammar Gaddafi was overthrown.

STS to FD:

3. In view of your observation that "In the end it is how you fight, as much as why you fight, that makes your cause good or bad," [*DU* p. 41] what are your thoughts on recent events in Tunisia, Egypt, Libya, and Syria? (We proudly notice that students were leading the way in many of these protests.)

FD to STS:

3. Recent events in the Middle East are too recent for me to pass any judgments. The effects of good or bad ways of fighting only become clear after a few years. My own judgment is that for students, to engage in passive resistance to governments is likely to be successful, and to engage in rioting and bloodshed is likely to be unsuccessful. But it will take time to find out whether that is true. In the meantime we can only admire the courage of those who passively resist. I happen to think it is very unwise for us to start a civil war as we are doing in Libya.

At the beginning of each semester when Professor Dyson is first introduced to a new class, I play a clip from a *Star Trek* episode[6] where the crew of the *Enterprise* comes across what they call a "Dyson sphere." In that episode a Dyson Sphere is a solid spherical shell enclosing a star, and the *Enterprise* crew envisions a vast number of people ("the equivalent of 250 million Class M planets") living on the shell's interior. In the real universe, a solid spherical shell surrounding a star would be unstable.[7] What Professor Dyson originally envisioned back in 1960[8] was an array of orbiting satellites that would surround the star, each satellite in its own orbit, receiving energy from the star and relaying it by electromagnetic waves to the host civilization.

STS to FD:

4. Do you think we humans will ever be able to come up with enough resources to build a Dyson Sphere?

FD to STS:

4. The "Dyson sphere" is a total misunderstanding of what I had in mind, but we will probably develop big settlements and industries in space, which will make grand-scale engineering projects feasible. The question is not whether these are technically possible, but whether we will have the motivation to carry them through. If you look at the Solar System as a whole, there is no lack of resources to do all kinds of crazy stuff. And some of the big enterprises may make sense. It all depends on human choices which we cannot predict.

STS to FD:

5. What is your most pressing fear, and your greatest hope?

Thank you for all you have done for us, through your books, articles, letters, and influence. Best wishes to you, to Imme, your children, and your grandchildren.

—STS class, Spring 2011

FD to STS:

5. Most pressing fear is still a war involving nuclear weapons. Greatest hope is a public campaign to eliminate nuclear weapons both unilaterally and multilaterally.

Sorry I am leaving for England and can give you only these brief answers.

Best wishes to you all, yours ever, Freeman Dyson

STS to FD:

July 16, 2012

Dear Professor Dyson,

It has been a long time since our Science, Technology, and Society class has contacted you. We did not want to make nuisances of ourselves, and we have a stockpile of questions from previous classes and your answers....

We have often wondered if easy access to essentially unlimited information via the Internet and Google has resulted in diminished interest among today's students about engaging an author personally... We have noticed that, among most students, as soon as they go outside the tiny screen comes out and they walk across campus with their heads down, looking at the small screen instead of the wide world all around them. Both the small screen's capabilities and the wide world deserve their respect and attention.

... If this course accomplishes nothing else, I hope it helps motivate students to ask questions—especially when understanding the questions may be more feasible than coming up with answers. As Walter Kaufmann observed, certainty should not be purchased at the price of honesty...[9]

After some discussion, here are the questions that the students from the Spring 2012 section wish to ask you. If you could find a few moments over the next few weeks to respond to them, we would be very grateful.... I should have sent them the day after the class was over, but have been distracted with getting back into our house after a tornado swept up our street in May 2011. A year later, the construction (with all its requisite decisions, and the moving itself) was winding down as the semester ended.

.....We understand that you will be one of the featured speakers at the Sigma Pi Sigma physics honor society's quadrennial conference, which occurs in Orlando this coming November....

On behalf of the 49 students in the Spring 2012 STS section, here are their questions:

1. We can presumably assume that science has provided great meaning in your life. But have the most meaningful things in your life come from science? Broadly speaking, in your view what makes life meaningful?

FD to STS:

16 July 2012

Dear Dwight,

Thank you for the message from your class and the two attached documents which I have printed out. I look forward to reading all of them and giving them the attention that they deserve. I will respond to the questions when I have time to think about them. I write now just to let you know that I received them and am grateful. Since the students are already gone, there is no reason for me to reply in a hurry.

You never told me that your house was demolished a year ago. Congratulations for surviving! That reminds me of the term that I spent in 1995 teaching at Gustavus Adolphus college in Minnesota. I arrived about six months after half the town and the campus were demolished by the biggest tornado in Minnesota history. It was a wonderful time to be there, with everyone helping each other to survive and sharing the hardships. I hope you and your neighbors are enjoying the same survivors' high after the disaster.

I am in California very busy with Jason work and more travels ahead. I will write again after we get home in August. The family is in good shape and so are we.

Yours ever, Freeman

I am proud to report that after our EF5 tornado passed through our town, everyone pulled together, everyone did whatever they could to help. The day after the tornado, as we were sifting through the rubble to see what could be salvaged, someone handed me a phone. A reporter from the metro newspaper asked for a statement. I told him that a tornado taking all the houses off their foundations in our neighborhood was not an "act of God;" it was an act of nature and we merely happened to be in the way. But, I said, if you are looking for an act of God, look to the many people who turned up to help us handle the debris, who are providing meals, offering places to stay, and so on.

FD to STS:

July 17, 2012

Dear Dwight,

Since your students have only three questions, I have time to answer them right away. I enjoyed your chapel talk very much, especially the song-lines which were new to me.[10] *I shall probably borrow your lines the next time I am talking to students.*

Here are the answers.

1. I have always said that the most important things in my life are in this order, family first, friends second, work third. And for me, work includes writing books as well as doing science. So altogether, science is only a small part of my life and not the most meaningful. Doing science for me is like playing a concert for a musician. It is exercising a God-given skill which I am happy to share with an audience. I do not care whether the science that I do is important. It gives me the same joy, whether it is important or not.

Yesterday something happened which is for me more meaningful than science. I spent the day with my 20-year-old grandson Randall whom I had not seen for two years. In two years, Randall changed from a shy and inarticulate teen-ager to a self-assured and thoughtful grown-up. Suddenly he is my friend and colleague and I enjoy listening to his stories. In spite of my distrust of higher education, I have to admit that three years of college have done him a lot of good. He is applying to medical school at UC San

Diego and has a good chance of being accepted. I am proud to be his grandfather.

STS to FD:

2. Were the successful flights of *Space Ship One*, and, more recently, a payload delivered to the International Space Station by Space-X, the kinds of alternatives to NASA that you hoped for when you wrote "Pilgrims, Saints, and Spacemen"? [*DU*, Ch. 11] Looking ahead, what should be the role of government-sponsored programs such as NASA?

FD to STS:

2. I am delighted that private companies are moving faster into the space business. The Falcon rocket is a big success. But this is very far from the sort of independent space ventures that I was writing about in "Pilgrims, Saints and Spacemen." The companies that are active in space today are still heavily dependent on the government. Without government funding they would not survive. For truly independent ventures to be possible, space operations must become enormously cheaper. I believe this will happen, but it will take a long time. Perhaps in a hundred years from now.

As this record of correspondence was being assembled, in 2022, several Space-X rocket flights have been completed, taking supplies and crew members to the International Space Station, and astro-tourists on sub-orbital flights. Professor Dyson's reply to Question 2 continues:

Meanwhile there are plenty of good things for NASA to do. NASA is already doing a splendid job with unmanned missions exploring the universe, and these will continue. The recent Kepler mission is huge success, discovering hundreds of planetary systems orbiting around other stars. The big problem for NASA is to find something exciting to do with manned missions. Manned missions are not needed for science. To have a meaningful program of manned missions, we have to think in centuries rather than in decades. Kennedy started NASA in the wrong direction when he set the aim of the Apollo program as putting a man on the Moon

> *and bringing him back in ten years. As a result, the program was not sustainable and turned out to be a dead end. Kennedy should have said, we start a modest and sustainable program that will bring people to live permanently on the Moon and planets in a hundred years. If he had said that, we would already be half-way there.*

Looking back at the international politics of the late 1950s and early 1960s, it is clear that President Kennedy's motivation for setting the goal that culminated in the Apollo program was less about the exploration of the universe and more about prestige in the politics of outshining the Soviet Union in order to win influence among the "unaligned" nations during the Cold War. Perhaps future generations will remember the accomplishments in technique and spirit of the Mercury, Gemini, and Apollo programs long after the political motivations for their timing has been forgotten.

STS to FD:

> 3. Since it is not clear today at whom the USA would shoot its stockpile of several thousand nuclear weapons, why do we keep so many of them? We understand that their maintenance costs tens of billions of dollars per year, while education (for example) seems to be perpetually under-funded. Wouldn't, say, a hundred nuclear weapons be enough? What is the role in all of this of the vested interests that President Eisenhower called the "military industrial complex"?[11]
>
> We hope you and your family members are all doing well....
>
> Best wishes and warm regards, STS class

FD to STS:

> *3. I agree that a hundred nuclear weapons are more than we need for any reasonable purpose. I believe strongly that we would be better off with zero, even if our enemies have more. My friends are all worrying about nuclear weapons in North Korea or Iran, when they should be worrying about our own weapons. Our own*

weapons are far more dangerous to us, being spread around the world in places where they might be captured or stolen. And the best way to get rid of the weapons is to do it unilaterally, as George Bush senior did in 1991, when he got rid of more than half our weapons. He got rid of all the weapons belonging to the army and the surface navy. Now we have only those belonging to the air force and the submarine navy. The military-industrial complex did not oppose Bush's decision. The soldiers and sailors in the army and surface navy were glad to be rid of the nukes, which got in the way when they had to fight real wars.[12]

There are of course big, vested interests which oppose getting rid of nukes. But the vested interests are mostly civilian rather than military. The politicians are generally worse than the soldiers. The soldiers know what war is like, and they know that nukes are not likely to be useful when they have to fight. The politicians mostly like nukes because nukes bring jobs to their districts. George Bush senior had the great advantage of being a right-wing Republican, so he was not afraid of the politicians.

To conclude, I wish you all a great future, whether you are Democrats or Republicans. Although I would never vote Republican, I have to admit that Republicans are not all bad. George Bush senior twice showed great wisdom, once when he got rid of more than half of our nukes, and once when he avoided occupying Baghdad after defeating the Iraqi army.

Yours ever, Freeman Dyson

In the spring of 2011 my wife and I took my father to London in honor of his 80th birthday. Since Professor Dyson grew up in London, I shared some scenes from that trip with him. His response sparkles with the enjoyment of life:

FD to DN:

31 August, 2012

Dear Dwight,

My wife and I made two trips to England this summer. We did not do as much as you and your father, but we also enjoyed walking in the streets and eating pannini in the Cafe Nero coffee-shops that have sprung up everywhere in recent years. I still feel at home there after an absence of sixty-five years. One trip was to celebrate my little diamond jubilee which luckily came three days before the Queen's big one. Mine was at the Royal Society in London where I have been a Fellow for sixty years. I felt like the Prodigal Son. They killed the fatted calf and gave us a friendly lunch on the roof of the Royal Society with a splendid view over central London.

Yours ever, Freeman

17 Living Through Four Revolutions

You say you want a revolution
Well, you know
We all want to change the world
You tell me that it's evolution
Well, you know
We all want to change the world...
—The Beatles, *"Revolution"*[1]

In September 2012, Nanyang Technological University in Singapore announced that it would host a conference in August 2013 to honor Professor Dyson on the occasion of his 90th birthday. In the meantime, Professor Dyson had accepted an invitation from the Sigma Pi Sigma physics honor society to be a plenary speaker at the society's 2012 Congress.[2] This Congress was held in Orlando, Florida in November. About 800 people were in attendance, some 600 of them physics students. The Congress theme was "Unifying Fields: Science Driving Innovation." As a celebrated visionary for innovation, Professor Dyson was scheduled to speak in the Saturday morning plenary session. Instead of swooping in on Saturday morning, he arrived on Thursday, the opening evening of the Congress. Walking into the convention center ever so nonchalantly, he was immediately surrounded by an enthusiastic crowd that included not only students, but also physics society executives and Nobel Laureates (Fig. 17.1).

This adulation, similar to what one might expect to see with the appearance of royalty or a rock star, continued throughout the entire conference. During every break, a long line immediately formed before Professor Dyson. Everyone wanted to shake his hand or take a photo or get his autograph—or all of the above (Figs. 17.2 and 17.3). He obliged everyone, turning no one away.

Fig. 17.1. Professor Dyson upon his arrival at the 2012 Sigma Pi Sigma Congress in Orlando, Florida, Nov. 8, 2012. He was surrounded by students before he could remove his jacket (DEN photo).

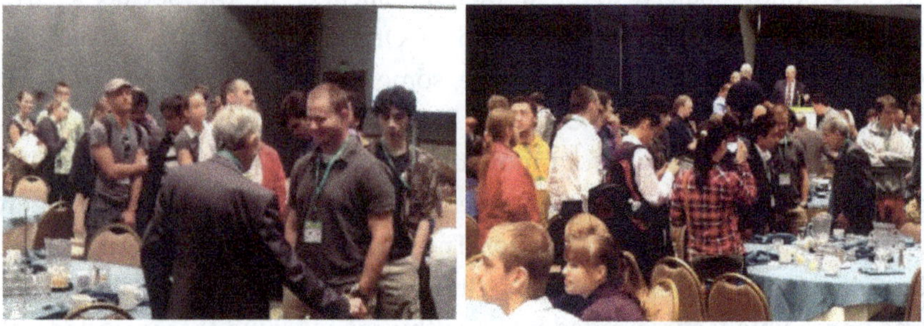

Figs. 17.2 & 17.3. Professor Dyson with crowds of students waiting to greet him at the 2012 Sigma Pi Sigma Congress, Orlando, Florida (DEN photos).

Friday morning featured an interactive workshop, "Connecting Scientists and Science Policy." It was led by David Mosher of the Congressional Budget Office and Anna Quider of the US State Department. With everyone seated in groups around tables, Mosher and Quider presented contemporary itemized Federal budget figures and asked our Congress attendees to recommend what we would cut from the budget if it had to be reduced by a specified number of billions of dollars.

Before the workshop began Professor Dyson quietly joined one of the tables, another workshop participant (Fig. 17.4).

Fig. 17.4. Professor Dyson with students and faculty in the federal budget workshop discussion (DEN photo).

While seated at the table with Professor Dyson before the workshop began, our conversation turned to some of the STS topics. I mentioned his passage in *DU* where he noted it is *how* one fights, besides *why* one fights, that can turn a good cause into a bad one. [*DU* pp. 40–41] A follow-up question mentioned the September 11, 2001 hijackers, asking Professor Dyson why, in his opinion, did they "hate the US so much" to carry out such an act. His answer was startling: He replied "I disagree." They did not follow through with that act out of hatred for the US, he said, but "because they loved each other." He had read diaries of the pairs of Japanese kamikaze flyers who resolutely set out on their one-way missions during the closing days of World War II.[3] Those young men went through with it, Professor Dyson recalled, not out of love for the Emperor or out of hatred for the United States, but because as kamikaze companions they found themselves in this grim situation *together*. Their assignment put them in the suicide plane, but their loyalty to one another enabled them to see their fatal missions through. (Incidentally, when Professor Dyson disagrees with you, he does it ever so politely.)

On Saturday morning Professor Dyson delivered his plenary talk. His abstract was published in the Congress program announcements:[4]

Living Through Four Revolutions

I had the luck to begin my professional career in 1945 when four radical scientific revolutions happened simultaneously. They were space, genomics, nuclear energy and computing. Looking back 67 years later, we can count up the score. One was a flop, one was a huge success, and two were half-and-half. That is not a bad score for scientific gambling.

Nuclear energy was a flop because nuclear energy never fulfilled our dreams of a cheap and safe and clean source of power. Computing was a success far beyond our dreams, becoming cheap and small and user-friendly and empowering ordinary people all over the world. Space was half-and-half, with unmanned missions a big success for science and manned missions a big flop for human adventure. Genomics was also half-and-half, a big success for science and a big flop for medicine. Space and genomics are wonderful tools for science but not doing much for human welfare.

It was inspiring to watch and listen as Professor Dyson delivered his speech, not only for its content, but also for his delivery. He used no notes. But the speech flowed beautifully from one point to another. Back in 2000 at the Templeton reception at the St. Albans School, I asked Imme if, when Professor Dyson writes his books, he has to revise the draft multiple times. No, she said, he thinks it all through carefully then writes it down, requiring few changes afterwards.

The long lines of admirers that clustered around Professor Dyson continued throughout the entire Sigma Pi Sigma Congress (Fig. 17.5).

Fig. 17.5. The crowds of students around Professor Dyson never ceased throughout the entire conference (DEN photo).

The conference ended on Saturday night. After 10 PM, when the convention center was practically deserted and the staff was dis-assembling the registration booth, Professor Dyson was nearby even then, talking with the last of the students who had lingered in the hope of having a word with him. He extended to everyone the graciousness he shows to my STS students. It was approaching midnight when the last of the students finally let him go. Of course, he was much younger then, a mere 88, going on a spry 89!

The next week found me back with the STS class.

STS to FD:

20 November 2012

Dear Professor Dyson,

It was good to see you at the Sigma Pi Sigma Congress in Orlando, and to talk with you. Your observations about the kamikaze pilots (with comparisons to the authors of the September 11 events) were a cause for reflection. Human beings will do for, or with, or to one another, out of loyalty, what they would never do alone. Perhaps together they become another person, and that is the person who does the deed, for heroism or for evil.... Sometimes the distinction between evil and heroism is ambiguous.

Your patience with the long lines of students at the meeting who wanted to meet you was impressive...You did not turn away a single student who wanted to speak with you.

...In our STS class discussions several questions to send you were proposed...

1. What is a non-science issue with which you have had to struggle, and what conclusions did you reach?

FD to STS:

25 November 2012

Dear Dwight,

Thank you for your friendly message. It was good to see you in Orlando. I forgot to ask you how you are recovering from the tornado, which hit you a hell of a lot harder than the hurricane hit us. I imagine you will still be repairing the damage for a long time. We have nothing to repair [from a recent hurricane] except a fallen tree and a big hole in the garden fence.

I am alone here in Princeton so it is a good time to answer your student's questions. The day after Thanksgiving my wife Imme nobly flew to England to spend a week with my sister Alice who is brain-damaged and ailing in an old people's home. I am selfishly staying home, with the excuse that I already made two visits this year. Imme is now at the end of her first day in Winchester, which is perpetually grey and usually raining at this time of year. Alice is always happy to see us, and talks coherently about the past, but she forgets almost everything of the present. Thank God she is not unhappy, but she is no longer the sister I have known for 85 years. I hope when my time comes my heart will quit before my brain.

Now it is time to begin with your questions.

1. What non-science issues have I had to struggle with? I have been so lucky in my personal life that I never had to struggle hard. Like many scientists, I reached a mid-life crisis at the age of forty when it was obvious that I was no longer smart enough to compete with the young people at the cutting edge of science. It was clear that I must find another line of work. At the Institute for Advanced Study where I had a permanent job, I was supposed to do research and not required to teach. So I decided to resign my position at the Institute and move to a university where I would spend the rest of my life as a teacher. I had good invitations from Yeshiva University in New York and at Northwestern University in Chicago. But when I discussed the possibility of moving with my family, they voted unanimously against it. They did not want to live in New York or in Chicago. I did not have the moral strength to go against the family wishes. So I stayed at the Institute and started a new career writing books for the general public. Writing books is my way of teaching. For the second half of my life, I was a writer rather than a scientist, doing research as a side-line, working on unfashionable problems which did not require me to be competitive. I found writing to be an enjoyable and rewarding activity, and through my books I came to know a much wider variety of people than I had known as a scientist. Now at age 88 I am reviewing other people's books instead of writing my own.

On an intellectual level, I had another struggle, adapting my socialist principles to a capitalist society. In England during World War Two, I lived in a socialist society that functioned well. That was perhaps the highest point that socialism ever reached. Money really did not matter. Everyone got the same rations of food and clothes and soap and other necessities. The rationed stuff was cheap, and there was nothing else to buy. Cars were not allowed any gasoline except for official business. It was a wonderful time to be a socialist, so long as the war lasted. Rich and poor people were all in the same boat. We tried to continue living the same way after the war ended. But gradually, as the rationing was abandoned and more things became available, money started to matter. Rich and poor became different. Inequalities became sharper. When I started to raise a family, I discovered that my socialist principles gave way to my responsibilities as a father. As a father, I needed money to take care of my wife and kids, and the more money the better. The theoretical idea of equality faded, as the kids needed a good home in a good neighborhood with good schools. The final blow to my socialist ideals came when my oldest daughter became a successful venture capitalist, using her wealth to rescue small companies and start new enterprises. She taught me that capitalism can be creative. She became a generous friend and role model to her numerous nephews and nieces.

STS to FD:

2. The Gospels tell us that when Pilate asked Jesus, "What is truth?"[5] Jesus gave no answer. If someone asks you, "What is truth?" how would you respond?

FD to STS:

2. What is truth? This is a question of words rather than substance. Jesus, according to John 18:37, said "To this end I was born, and for this cause I came into the world, that I should bear witness unto the truth. Every one that is of the truth heareth my voice." And Pilate said, "What is truth?" and did not stay for an answer. Pilate had no idea what Jesus had in mind. As usual, Jesus did not explain himself. I imagine that what he meant by truth was a clear vision of things as they are. But that is my own interpretation, and Jesus may have had a different notion of truth. Pilate's previous question was, "Art thou a king?" Jesus said, "My kingdom is not of this world." So his notion of truth may have been a vision of a different world.

Each of us has a personal set of beliefs that we consider to be the truth. For me, truth is not absolute or permanent. It changes as we learn or as we forget. Truth is a convenient way of organizing our fragmentary understanding of the world. The beautiful thing about science is that almost everything we believe turns out to be wrong. Science is perpetual learning, perpetual bumping into fresh surprises.

Time to go home for supper. I will finish this on Sunday.

Professor Dyson's comment that in science, "almost everything we believe turns out to be wrong," generated some class discussion. Some ideas (e.g., the geocentric model of the universe) are discarded when the frontier of science expands. Other concepts that were useful in their original domains are not discarded when a more comprehensive set of ideas emerge; they are understood to be special or limiting cases embedded in the more comprehensive theory. An example would be Newton's law of universal gravitation that was supplanted by Einstein's general theory of relativity. While Newton did not have the final story on gravitation—and in that ultimate sense Newtonian gravitation is wrong—nevertheless the Newtonian paradigm survives as a special case and remains useful in its original domain of applicability. One needs General Relativity to understand black holes, but when navigating astronauts to the Moon and back, Newtonian gravitation theory is sufficiently accurate.

STS to FD:

> 3. What is the solution to climate change for this century, and in the long term?

FD to STS:

> *3. What is the solution to climate change, for this century and in the long term? Answer, climate has always been changing and always will be changing. We do not understand why climate changes and we cannot predict it. So the best we can do is to adapt to change as it happens. There is no ideal solution to the problem. One place where climate change is most extreme is Illulisat, a place in Greenland where Al Gore goes to take his spectacular pictures of melting glaciers and disintegrating icebergs.*[6]
>
> *I also went to Illulisat and saw everywhere the evidence of rapid warming. I talked with the people who live there and they love the warming. They hope it continues. It makes their lives much easier. In the old days when it was colder, they lived by fishing in the ocean, and one third of all their young men died at sea. Now they stay on land and grow vegetables and build hotels for all the tourists who come to take pictures of the melting ice. The fishing boats are now taking tourists for trips to the neighboring islands. Tourists pay much better than fish. There is no doubt that climate change is good for Illulisat.*
>
> *The big unanswered question is whether climate change is natural or caused by human activities. Many people believe that it is mainly caused by human activities. I believe it is mainly natural. It will take a long time before we know who is right. In the meantime, China and India will continue to burn large quantities of coal in order to become prosperous modern societies. Coal-burning is*

certainly bad for the environment. Fortunately we recently discovered ways to drill for shale gas which is a much cleaner substitute for coal. There are big reserves of shale gas well distributed over the earth, in China, in India, in the USA and in Europe. In the USA, shale gas is already replacing coal.

I would say, for this century the continued development of shale gas will counteract the bad effects of coal burning. Then, after a hundred or two hundred years, when the shale gas runs out, we will probably have cheap and efficient ways of using solar energy. In the long run, solar energy is the way to go, whether or not it helps to mitigate the effects of climate change. But to spend big money now, subsidizing expensive ways of using solar energy, makes no sense.

STS to FD:

4. If you wanted to begin in a non-science career now, what would it be, and why?

FD to STS:

4. If I wanted to begin in a non-science career now, what would it be, and why? The main requirement for a successful and useful career is a high level of skill. You must be highly skilled to do any job well. I have two basic skills, calculating and writing. So my choice of a career is either to do mathematical calculations or to do literary exercises. In practical terms, that means either science or journalism. If I would begin a non-science career now, it would be a writing career. I would begin with a job as a journalist or a teacher, with the hope of writing a best-seller and becoming independent. My daughter Esther began her professional career as a journalist, took a job with Forbes which is a business magazine, then decided that it would be more fun to be a player than a spectator, and ended by becoming a successful venture capitalist. If I were young and beginning a non-science career, I would start the same way as Esther did and look for unexpected opportunities.

STS to FD:

5. What advice do you have for this generation of students? Thank you Professor Dyson. Have a great Thanksgiving!

-Fall 2012 STS class

FD to STS:

5. What advice do I have for this generation of students? I would say the same thing that my daughter Esther said when she was giving a commencement address to the graduating class at Carleton College. Worry about your fourth job, not your first. The first job is just to get your foot in the door. Don't expect to enjoy it. Use your first job as a training to find out what you can do. The first job also gives you a better bargaining position when you are looking for the second job. And so it goes on. If you are still unhappy with your fourth job, then you have a problem.

I do not have much more to say. Each of you is different from the others and has different needs. Some like stability. Others like adventure. If you like stability, choose a safe career path as a teacher or a nurse or a police officer. If you like adventure, be ready to grab at any opportunity that comes by. My daughter Esther, who likes adventure, has a motto which appears at the end of her E-mail messages, "Always make fresh mistakes." Do not do the same dumb thing twice. That is good advice for all of you.

Happy Christmas and New Year to all of you. Yours ever, Freeman Dyson

18 The Water Meadows

Ah, not to be cut off,
Not through the slightest partition
shut out from the law of the stars.
The inner—what is it? if not the intensified sky
Hurled through with birds
and deep with the winds of homecoming.
—Ranier Maria Rilke[1]

The Spring 2013 semester's STS class was still basking in the afterglow of the 2012 Sigma Pi Sigma Congress.

STS to FD:

26 April 2013

Dear Professor Dyson,

It was so good to see you and have a chance to converse with you in Orlando last November. I hope you had a quiet, peaceful remainder of the weekend when you returned home. I would like to thank you, on behalf of all the students, for participating in the Sigma Pi Sigma 2012 Congress. Your presence contributed much to making it a meaningful event.

Once again this semester, the students in our Science, Technology, and Society class...would like to ask you some questions. We know that you are very busy and do not want to trade on your generosity with your time that you exhibited in Orlando....

1. Humanity is currently approaching a Type 1 [*DU* p. 212] civilization, where we control (or think we control) the resources of a planet. In the long-term future, we may advance to a Type 2 and eventually to a Type 3 civilization, assuming humans do not become extinct in the meantime. Professor Dyson, do you think humanity will survive these transitions? To become that advanced as a species, we must

be extremely organized as a civilization. Perhaps a society in possession of the qualities necessary to be a Type 2 or 3 civilization would be similar to a colony of ants or bees, requiring a centralized figure (the queen bee) that directs the activities of the workers. For human beings to be at least as efficient as bees and ants we may have to adopt a universal "hive mentality." This mentality would be beneficial to the whole for preserving and advancing our species. But at what cost? Would humanity benefit overall if the majority of human beings (the proletariat) became worker bees while the elite became the queens? How much of our humanity would we have to sacrifice for this efficiency? It is our humanity that allows us to peer into the depths of the universe via intuition, reasoning, and intellect. It is our humanity that leads to scientific discoveries. It is our humanity that creates art, music, and literature. Could our species continue evolving towards Type 2 and 3 civilizations, while also retaining the qualities that have allowed us to flourish? Or will the beautiful gift of intellect, and its faithful tool, science, be used against the ones it was made to protect and serve?

FD to STS:

30 April 2013

Dear Dwight,

Thanks again for the delightful week-end in Orlando. I certainly enjoyed the kids at least as much as they enjoyed me.

This time you have fewer questions than usual and I can answer them quickly. Here are my immediate reactions.

1. The whole point of moving out into the universe is that the universe is big. Going out into space means that it is easy to hide, easy to declare independence, easy to stop paying taxes, easy to avoid the neighbors, easy to live differently. Any sort of central authority is likely to be weak and temporary. The price we pay for this freedom is speciation. Like other successful species, we shall probably diversify into many different species adapted to local environments. Astronomical distances will make it possible for many diverse species to coexist more or less peacefully. Some of our descendants may develop into bee-hive societies as you describe them, but more of them are likely to become loosely organized communities or solitary wanderers, just as it has happened in the ecology of apes and insects on this planet. The bee-hives may defend their territory fiercely, but there will be plenty of space outside their reach. That is my view of the future. I may be wrong, but I think your vision of a bee-hive society is made much less likely if we are spread out in the vastness of space. Everything you say about art, music, literature and science points in the same direction. Diversity and creativity go well together with huge distances under an open sky.

STS to FD:

We hear a lot about armed drone aircraft being used in Pakistan and Afghanistan. What are the effects of these drones on the USA's relations abroad? The drones can hit small targets with relative surgical precision (compared to carpet-bombing or atomic bombs or chemical and biological weapons), but they also suddenly strike a village in Pakistan from above, flown by someone sitting in New Mexico. In what sense do drones contribute to further "making evil anonymous," [*DU* Ch. 3] and in what sense do they perform a genuine service? We hear that some cities are considering the use of drone aircraft in traffic enforcement. Are we right to be concerned?

FD to STS:

2. The answer here is simple. We are right to be concerned. Drones are blurring the boundary between war and peace. They raise big and difficult ethical dilemmas. I feel some personal responsibility for this development, because of the company I helped found and used to work for, General Atomic, is now the world's biggest manufacturer of drones. Last summer I visited their factory in California, an enormous building that helps to revive the economy of California with well-paid jobs. Twenty years ago, the company decided that the nuclear business was a loser and they had to find a new line of work. They made the decision to start a new business with drones, and I gave it my blessing. At that time we were thinking of drones as unarmed spy-planes, and it seemed like a good idea to take the pilot off the plane. So the company designed and built the Predator A, the spy plane. But then somebody in the company had the bright idea of putting missiles on board, and it became the Predator B, designed for killing people. Now the Predator B is the main production line and the company is making bigger profits than ever before. It should have been obvious to me that this would happen, but it came to me as a surprise. Visiting the factory last summer, I found it very scary. Such a huge and profitable business. There is no way I can imagine it being stopped. That is the future, whether we like it or not.

Whether the drones are good or bad depends on the alternative. If the alternative is to send an army of 50,000 soldiers with boots on the ground, then drones are good. Drones kill a few people and create a lot of resentment, but an army kills far more and creates far more resentment. If the alternative to drones is to catch the bad guys with old-fashioned police work, then drones are bad. In the real world, it is hard to know where to draw the line. It is not surprising that Obama loves drones. Drones allow him to keep on killing bad guys while he is bringing armies home. Drones are dangerous to our future because they are acting secretly and with no legal restraints. The job for your generation is to learn how to live with drones in a dangerous world.

STS to FD:

3. Which of your achievements has brought you the most personal fulfillment?

Thank you again for being part of our class. We have enjoyed reviewing correspondence that you carried on with past STS classes, and appreciate your contributions to ours.

-Spring 2013 STS class

FD to STS:

3. The achievement that brought the most joy is raising six kids who all turned out well, successful in a variety of careers and raising families of their own. After that, the next achievement was

> *solving a variety of puzzles in mathematics and physics, none of them very important but all of them beautiful. After that, teaching and writing books for students like you. Thank you for continuing to ask questions.*
>
> *Yours ever, Freeman Dyson*

Few, if anyone, would agree that Professor Dyson's solutions to "puzzles" are unimportant. Not only does the physics community recognize Professor Dyson's contribution to Quantum Electrodynamics as deserving the Nobel Prize,[2] but he was recognized with essentially every Award or Medal or Prize that science societies around the world offer.[3] In August 2013, the Institute of Advanced Studies at Nanyang Technological University in Singapore held the conference (Fig. 18.1)

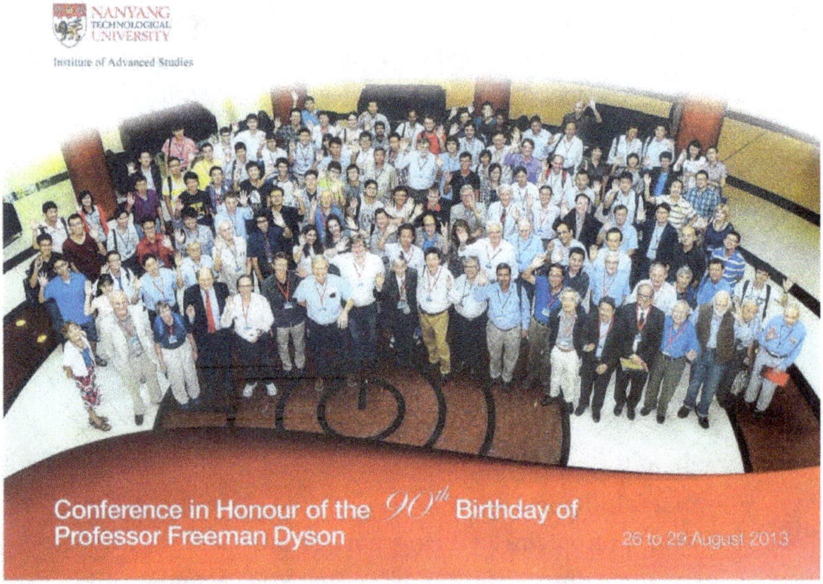

Fig. 18.1. Participants in the 2013 conference honoring Professor Dyson's 90th birthday. Professor Dyson stands in the middle of the front row; Imme stands in the front row on the far left (Conference photo).

> ...in honour of the 90th birthday of Professor Freeman Dyson, a great physicist, thinker and futurist, whose impact goes beyond quantum field theory and statistical physics,

into other fields of mathematics and physics, literary and public policy.

When the call for the conference went out, there was an overwhelming response from scientists everywhere. Many of his former colleagues....agreed to make presentations at the conference...

Many others who were of a different generation also responded to the call...[4]

Professor Dyson delivered the opening presentation: "Is the Graviton Detectable?" (Fig. 18.2).[5] At the age of ninety, Professor Dyson was once again bringing cutting-edge physics issues into sharp focus.

Fig. 18.2. Professor Dyson delivering the opening address, "Is the Graviton Detectable?" in the Singapore conference (DEN photo).

...This talk is concerned with a different question, whether it is in principle possible to detect individual gravitons, in other words, whether it is possible to detect the quantization of the gravitational field...

...We have examined three possible kinds of graviton detector with increasingly uncertain results...

In this talk I am not advocating any particular theory of a classical gravitational field existing in an otherwise quantum-mechanical world. I am raising three separate questions. I am asking whether either one of three theoretical hypotheses may be experimentally testable. One hypothesis is that gravity is a quantum field and gravitons exist as free particles. A second hypothesis is that gravity is a quantum field but gravitons exist only as confined particles, like quarks, hidden inside composite structures which we observe as classical gravitational fields. The third hypothesis is that gravity is a statistical concept like entropy or temperature, only defined for gravitational effects of matter in bulk and not for effects of individual elementary particles…

The conclusion of the analysis is that we are a long way from settling the question of whether gravitons exist. But the question whether gravitons are in principle detectable is also interesting and may be easier to decide.

Fig. 18.3. Professor Dyson about to cut his birthday cake at the banquet of the Singapore conference (DEN photo).

At the Tuesday evening plenary session the featured speaker, a well-known physicist, used much of the speech to launch a mocking tirade against religion in general. Offering no caveats or nuances, the speaker painted with a broad brush the notion that all religions are folly. Meanwhile, Professor Dyson, who famously wrote about the "two windows," who delivered a moving and balanced speech on science and religion to the Conference of Catholic Bishops, who received the 2000 Templeton Prize for Progress in Religion, maintained a dignified silence. Professor Dyson's polite silence said more than words could have said.

Near the end of the Singapore conference, while waiting for a session to begin, in a conversation with Professor Dyson I mentioned having received the news, through the annual Dyson family's New Year letter, about the passing of his sister Alice earlier in the year. I expressed condolences from myself and on behalf of the STS students. He related the closeness that he and Alice enjoyed throughout their lives. Alice passed away at age 92 on February 13, 2013, in Winchester, "the town where we spent our childhood and where she spent most of her life as a medical social worker," as the January 2013 Dyson Family Chronicle explained. The Chronicle gave details of Alice's final months:[6]

> *Imme flew over to be with her for the last days. Imme was rubbing her back to let her know that she was not alone when she died. In February Imme and I flew over for the funeral at the Catholic church of St. Peter. A crowd of about 200 filled the church. Alice had a multitude of friends in the town, and family came from all over England. Three of our children, Esther, George and Rebecca, flew over from America. In April, Imme flew over for the last time, to scatter Alice's ashes in the Water Meadows where she loved to go walking with her dog. Alice was my closest friend and a beloved aunt to our children. For three years, Imme has been flying over frequently to take care of her, to arrange for the sale of her house and for her to move into the old people's home where she spent her final months and quietly died.*

Upon returning home from the Nanyang conference, the beginning of the fall semester was imminent. Within the first week of school, still buoyant from the conference, I shared the enthusiasm with this new class and wrote about it to Professor Dyson.

DN to FD:

4 September 2013

Dear Professor Dyson,

It was so good to see you and Mrs. Dyson again. Thank you for the times we spent in conversation. I deeply appreciate your friendship.

Last night I met for the first time this semester's STS class. There are 50 students enrolled. When the students see their textbook author not as a faceless "they" but as a friendly person who has grown children and 16 lovely grandchildren, their interaction with the text and subject matter takes on a deeper level of meaning. It's a joy to see.

...Thank you for helping me teach the course. I have grown in the process, and have found it very fulfilling. I came here to teach physics, and enjoy that immensely, but the STS adventures may turn out in the long run to be the more important....

Attached is a copy of the talk as I gave it last Thursday[7] [in Singapore].... Also I thought it was such a lovely moment when the gentleman from Bulgaria came up to you and told you about his encounter with the Thompson train station.[8] He and you had a short conversation in honor of Frank Thompson. I took a photo to commemorate that moment... (Fig. 18.4) I was glad to be part of the company that gathered at NTU to honor you....

Fig. 18.4. Professor Dyson and a gentleman from Bulgaria discussing Frank Thompson [DU Ch. 4] at the Dyson Conference, Nanyang Technological Institute, Singapore, August 2013 (DEN photo).

In the September 4 letter I informed Professor Dyson of another poll we conducted in class regarding the three options he had outlined in his Gustavus Adolphus Seminar. He replied at once:

FD to DN:

> *14 September 2013*
>
> *Dear Dwight,*
>
> *This is just to say thank you for the package of correspondence with your students that arrived yesterday....*
>
> *I was very much interested in the result (3, 37, 2) of your vote on genetic manipulation of babies.... There is a real difference between Oklahoma and Minnesota.*
>
> *Thank you also for sending the picture of the Bulgarian gentleman....*
>
> *As always, I shall be glad to hear from the students. Yours ever, Freeman*

We obliged the following December.

STS to FD:

4 December 2013

Dear Professor Dyson,

On behalf of the Fall 2013 semester Science, Technology, & Society class, we thank you once again for being part of our lives through *Disturbing the Universe*... Several questions were proposed and as a class we settled on these. We would be very grateful if you have a few moments to respond to some of them.

1. This semester we looked at the various options for automotive power, such as gasoline and diesel, electric cars, hybrids, steam cars, cars that run on compressed air or flywheels, and even solar powered cars (e.g., entries in the annual Sunrayce[9]). Your experience with Project Orion [*DU* Ch. 10] suggests that you would take seriously proposals for unconventional propulsion systems intended for everyday ground transportation. If the USA were going to launch an ambitious organized program (e.g., with federal funding) to develop alternative energy sources for ground transportation, what recommendations would you make for prioritizing our investment?

FD to STS:

6 December 2013

Dear Dwight,

Thanks to you and the students for the questions. I am glad to see that they are mostly new and give me a chance to say something different.

1. In general I do not think government programs to develop new kinds of cars or trains are a good idea. What we need is small risky ventures that can be allowed to fail,[10] *and government is not good at taking risks. Private ventures do the job better. Government programs are needed when the job is too big for a private venture, and then there is a danger that it becomes too big to fail, and it is kept alive for political reasons when it is really a loser. I see only one good candidate for a government program in this area, and that is the fast train project of Gerard O'Neill. He called it VSE for*

Velocity, Silence, Efficiency. The project died when O'Neill died in 1992. I believe Elon Musk has been promoting a similar project quite recently.

The idea of VSE is that any new ground transportation system must beat the existing system by a factor of ten to capture the market. So VSE is designed to beat the roads and airlines by a factor ten in velocity, a factor ten in silence, and a factor ten in efficiency. The fast trains are not really trains but little capsules traveling independently, each carrying six passengers with luggage through steel tubes. The tubes are standard gas pipe-line tubes with 8-foot diameter and are quite cheap. Inside the tube is almost a vacuum, with just enough air to conduct heat but not enough to create drag. The essential idea that makes the system efficient is that every trip is non-stop. There are no big stations like airports to delay travelers. There is a small station underneath every shopping-mall parking-lot. You park the car and get into the capsule and punch in your destination, and off she goes, maximum coast-to-coast travel time 50 minutes. The whole system operates like a telephone network, with a network of pipes and capsules programmed to avoid collisions. That was O'Neill's dream. When he was dying of leukemia, he gave me his model of the system to take to Washington. I had a meeting with the high-up people at the Department of Energy and showed them the model and explained how it would work. Of course, we all knew that the project could not survive without O'Neill's driving force to push it along. When he died, the project died too. Perhaps, within your lifetimes, it could be resurrected.

STS to FD:

2. Recent disclosures about data being collected by the National Security Administration, along with the growing use of surveillance cameras, data mining, and drone aircraft, raises interesting dilemmas in the interactions of technology with society. What moral principles, translated into policies and laws, should guide our decisions on these issues?

FD to STS:

2. There are two separate issues here. One is the collection of personal information about people. The other is the use of drones to kill people. Both are serious problems. The use of drones to kill people is a more urgent problem. It blurs the distinction between war and peace. If we persist in killing people this way, we will never have any real peace. The only solution that I can see to this problem is to stop killing people by executive action, to go back to the old rule of law, that you kill people only by legal process with a fair and open trial.

The collection of information is a more complicated problem. There are many good reasons to collect information. The problem now is that it is cheaper to store information than to destroy it, so huge amounts of information are stored and little is destroyed. The problem is to avoid the abuse of this information by vindictive people and governments. In my opinion the most urgent need is to get rid of secrecy as far as possible. The worst abuses happen when information is kept secret. Most of the secrecy is harmful and unnecessary. To me, anyone who breaks the rule of secrecy and brings the abuses out into the open is a hero. To punish whistle-blowers is a clear sign of tyrannical government.

STS to FD:

3. For estimating the number of communicative civilizations per galaxy, the numbers that go into the Drake equation[11] are so uncertain as to make any conclusions meaningless. You made this point in the chapter "Extraterrestrials" when you wrote, "I reject as worthless all attempts to calculate from theoretical principles the frequency of occurrence of intelligent life forms in the universe." [*DU* p. 209] That said, what does your intuition tell you? Given that the Kepler satellite has found at least one planet around every star it has surveyed so far, which outcome would be more difficult to understand: That life and intelligence is rare and perhaps unique in the universe, or that life and intelligence occur with a non-negligible frequency? In other words, if you were going to bet on the existence of other intelligent life in our

part of the galaxy close enough that we could exchange radio signals on a timescale within the scope of a civilization's lifetime, how would you bet?

FD to STS:

3. There are two possibilities about the origin of life.[12] *Either the origin of life is a lucky chance with very low probability, or it is a routine event with reasonable probability. So long as we do not understand the details of the process of originating life, both possibilities are open and it does not make sense to guess which is more likely. If I had to bet on finding evidence of life, I would certainly bet on the positive side, because a bet on the negative side could never win. That does not mean that a positive result is more likely.*

STS to FD:

4. You have traveled over so much of the world, visited so many countries, and have friends in so many cultures, may we ask what are your favorite places, and why?

FD to STS:

4. I do not have a favorite place, because each country has unique qualities, and each is enjoyable in a unique way. The best thing about this planet is that different cultures and different ways of living still survive. I will mention just a few places that I found delightful. The place that I visited most recently is Singapore. That is a wonderful place because it is a socialist welfare state with a flourishing capitalist economy. It combines two very different heritages, the Chinese heritage of capitalist work ethic, and the British heritage of socialist government. It succeeded in preserving the best bits of both heritages. It is a living proof that socialism and capitalism can work together.

Another wonderful place is Switzerland, which has a big diversity of cultures and languages in one small country. I lived in Zurich which is an old-fashioned city with electric street-cars and clean streets. When our first baby was due to be born, my wife arrived at the hospital almost too late, with labor already started.

The guard at the door would not let us in until we had given him two names for the baby, one for a girl and one for a boy. In Switzerland you are not allowed to be born without a name. We had to decide quickly. It was Oliver or Esther, and she turned out to be Esther.

Another place that we loved was the United States of sixty years ago, when we came to live at Ithaca in New York State. In those days, the United States was friendlier than it is now, students did not have cars, and we all helped each other to survive in the snowy winters. We were not afraid to hitch-hike or to pick up hitchhikers on the highways. Nobody was homeless and nobody kept their house-doors locked. Rich people were not so rich and poor people were not so poor as they are today. The war-time ethic of sharing the hardships still prevailed. The biggest need of this country today is to restore the ethic of sharing hardships. We have a lot to learn from other countries such as Singapore and Switzerland.

STS to FD:

5. In less than two weeks you will have completed 90 laps around the sun. For this we congratulate you! You are in excellent health, you still lead a robust and active life, and your services are still in demand. You are inspiring! With the view from 90, what advice do you have for university students in their early 20s, which would help them live a life not only as long, but even more important, is filled with meaning and fulfillment?

Thank you once again for being part of our course and our lives. Since 1986 some three thousand students have read *Disturbing the Universe* as a text for STS. That represents an enormous influence. We are grateful.

We hope that you and your family had a joyous Thanksgiving, and are looking forward to happy gatherings over the upcoming holidays. Best wishes to Mrs. Dyson, to your son and daughters and their families, and enjoy those wonderful grandchildren!

Warm regards, STS class of Fall 2013

FD to STS:

5. Thank you for your congratulations, but being ninety has not made me any wiser than I was before. My advice to you young people is still the same. Avoid making choices too soon, be ready to grab at unexpected chances when they appear, work hard at acquiring skills like programming computers or writing clear English that are always useful, make the best of whatever job you are doing, be ready to switch careers when necessary, be a good team player, and try to leave the world a better place than you found it. I like to repeat the advice that our daughter Esther puts on her E-mails: Always make new mistakes.

Happy New Year to all of you,

Yours ever, Freeman

19 Society and Sanity

What keeps the world from reverting to the Neanderthal with each generation is the continuing, ongoing mythos, transformed into logos but still mythos, the huge body of common knowledge that unites our minds as cells are united in the body of a man... To go outside the mythos is to become insane...

—Robert Pirsig[1]

STS to FD:

24 April 2014

Dear Professor Dyson,

We hope the first half of your 90th year has been filled with joy, time with family, many conversations with grandchildren, interesting work to do, and new adventures.

Once again our STS class risks wearing out our welcome by sending you some questions. If you have a few moments to answer some of them, we would be grateful....Each question will be followed by notes about class discussion during the semester that motivated the question.

1. If you could change anything about K-12 science education, what would it be?

Most of the students in this semester's STS class came up through the standardized testing that steers public K-12 education today. In class we discussed how many parents, students, and teachers are frustrated by the fact that the curriculum is not allowed to follow the children's insatiable curiosity. Investigations that could follow from questions in a third-grader's mind (e.g., "What holds the moon up when it's on nothing?" "Why are people different?") are, as the teachers say, "not on the test." The teachers unfortunately have no time to spare for going on interesting tangents, even though that would stoke pupil interest, make learning fun, and promote education appreciation even higher in the long run. The teachers' job security, student promotion to the next grade, and district funding are all tied to the test scores. The theme that initiated this discussion was the "six faces of science" that you articulated in your AAPT speech "To Teach or Not To Teach."[2] The ugly face of "science presented as a rigid authoritarian discipline" generalizes to a lot of disciplines as delivered in K-12.

FD to STS:

28 April 2014

Dear Dwight and Students,

Thank you very much for your questions. These are good questions and the discussions that came along with them are very helpful. You have already gone deeply into each of them. I will disagree with some of your opinions, but I do not expect to change your minds.

1. What would I change in K-12 education?

We all hate the tyranny of the testing system in American schools, but the system has one important virtue. It is fair. I happen to live in Princeton, New Jersey, where the Educational Testing Service (ETS) produces and administers the tests. I remember vividly a day when the Minister of Education of the Soviet Union came to visit Princeton. He visited three institutions, the Institute for Advanced Study where I work, the University nearby, and ETS. He was not much interested in the Institute and the University, but he was intensely interested in ETS. He said the Soviet Union had good Institutes and Universities but had nothing like ETS. He said the education system in the Soviet Union was grossly unfair. To get into a good university and have a chance of a good career, a child must have parents with good connections. Success depended on good connections, not on hard work. He said he would try to introduce something like ETS in the Soviet Union. Of course this never happened. The education system in Russia today is as unfair as it was in the Soviet Union in the old days.

I grew up in England where the system was unfair in a different way. I got a free education in an excellent private high-school. To get the free education beginning at age 12, I had to sit for three days taking ten written examinations in ten subjects including Latin and Greek. The examinations were intensely competitive. Only the top 13 out of hundreds of boys got in. Girls were not invited to compete. In theory the system was supposed to be fair because rich and poor boys were treated equally. In reality it was grossly unfair because only upper-class kids had a chance of being prepared with knowledge of Latin and Greek. The boys who got in were all like me, sons of well-educated upper-class families.

When I came to America and saw my kids taking the ETS tests in American public schools, I was delighted to see that the tests were fair. Also the kids did not need to spend much time preparing. The tests were rather trivial, being designed for average kids and not for the intellectual elite. I felt that the tests did less harm than the elitist British system or the corrupt Russian system. Now that my grandchildren are in American schools, I still feel the same way.

So finally I must answer your question. The main thing I would change in the American system is the number of hours that the kids spend in class. In my last year in the English high-school I spent seven hours a week in class. The rest of the time was free, for me and for the teachers, so that I could get an education as I chose and they could give me help when I needed it. The school had a good library, a music building with practice-rooms, a museum, a carpentry-shop and a book-binding shop, not to mention football and cricket fields. Sitting in class all day is not the best way to learn or to teach. If the kids and the teachers had more free time, the tyranny of the testing system would not be so harmful.

William Blake had something to say about childhood freedom on the one hand and, when things go badly, the times when school becomes dismal for those kids in the class:

The School-Boy

I love to rise in a summer morn,
When the birds sing on every tree;
The distant huntsman winds his horn,
And the sky-lark sings with me.
O! what sweet company.

> But to go to school in a summer morn,
> O! it drives all joy away;
> Under a cruel eye outworn,
> The little ones spend the day,
> In sighing and dismay....
>
> —William Blake, *Songs of Experience* excerpt (1794)

In a typical STS class, about a quarter to one-third of the students are Early Childhood or Elementary Education majors. They want with all their hearts to be teachers who create moments of joyous epiphany in their pupil's experience. At this point in their lives they are filled with passion for the education of children. Perhaps our correspondence with Professor Dyson will enable them to avoid the discouragement and burnout that comes from bureaucracy grown stifling, litigation grown too threatening, and state legislators grown too dictatorial.

STS to FD:

> 2. What social reforms will be necessary if nuclear energy turns out to be our only long-term energy source that can provide the quantity of energy needed to maintain national and global economies?
>
> In class we discussed environmental sustainability issues. For example, at present rates of consumption the USA goes through 8 billion barrels of oil (the reserve thought to be under the Arctic National Wildlife Refuge) about every 400 days. Clearly our dependence on fossil fuel is not sustainable. The business-driven response has been to look for more fossil fuel, at least for the time being (the earthquake frequency in Oklahoma has jumped dramatically since fracking began). Meanwhile, wind power is very controversial in this part of the country, as windmills seem to be popping up everywhere, resulting in citizen's committees, town meetings, and passionate letters to the editors of newspapers. Everyone wants cheap electricity, but no one wants a windmill or a nuclear reactor in their back yard. Curtailing our consumption of electricity, or pricing it to

reflect its true costs (including, for example, light pollution) does not seem to be in the cards anytime soon. Everyone wants cheap power but no one wants the costs to the landscape that go with it. In the long view we can't have it both ways. These discussions occurred under the motif of one of your lines from the chapter "The Greening of the Galaxy:" "Sanity is, in its essence, nothing more than the ability to live in harmony with nature's laws." [*DU* p. 237] By that definition, our society is insane.

FD to STS:

2. What social reforms will be necessary to provide us with a permanent and ample supply of energy? Here I disagree strongly with your discussion of the energy problem. I disagree particularly with your last sentence, where you say our society is insane. On the contrary, I think we are handling the energy problem much better than we are handling other problems such as poverty and inequality and gun-violence and education and public health. The obsession with energy is distracting attention from these more serious problems.

As I see it, nuclear energy is unimportant, a minor player in the energy game, not as good as its advocates claim, not as bad as its enemies claim. There are two major players, fossil fuels and solar energy. Fossil fuels are ample for at least the next hundred years and are allowing China and India to become rich. Solar energy is enormously abundant and will give us a permanent supply of energy as soon as we develop the technology to use it cheaply. I would be surprised if it takes as long as a hundred years to make solar energy cheap and available to everyone.

So my answer to your question is, no social reforms are needed to deal wisely with energy. Social reforms are needed to deal with the more serious problems, especially with inequality. Political actions to make fossil fuels more expensive make inequality worse.

Here again we were humbled by the words of our mentor who shows us a perspective larger than we considered. In a society where social justice prevails, where human dignity is respected by all, where proposals gain traction in proportion to how well they build community—in such a society, problems like sustainable sources of energy would be straightforward to solve!

In *The Scientist as Rebel* Professor Dyson wrote,[3]

> *If we can agree with Thomas Jefferson that these truths are self-evident, that all men are created equal, that they are endowed with certain inalienable rights, that among these are life, liberty, and the pursuit of happiness, then it should also be self-evident that the abandonment of millions of people in modern societies to unemployment and destitution is a worse defilement of the earth than nuclear power stations.*

When the next question was composed, Russia had recently invaded the Crimean Peninsula in order to take it by force from Ukraine.

STS to FD:

3. What do you make of the way Russia is treating Ukraine? What would you like to see happen there? What should be the response of the US and NATO to this situation?

With your experience at the Arms Control and Disarmament Agency [*DU* Chs. 12 & 13], your work with the Jasons, and your knowledge of Russian history, language, and culture, we thought you might have some needed perspectives on current events going on between Russia and Ukraine... If Ukraine or part of it genuinely wants to join Russia, there should be a way to facilitate that transition without violating Ukraine's constitution, international law, or human decency. On the other hand, we recall that Hitler annexed parts of Europe before WW II,.... saying to the world that Germany was merely protecting Germans. That did not turn out well. The motivation behind this question occurred after our discussions of the history of atomic bombs and the Cold War. Since President Putin is a former KGB officer, perhaps he still views the world through Cold War spectacles. We read in class one of your letters where you said our goal should not be to destroy our enemies, but turn them into friends.

FD to STS:

3. What do I make of the way Russia is treating Ukraine? I know very little about Ukraine, but it seems extremely unwise for the USA to become involved in a power-struggle in a place where we have no power. There have been two occasions when countries in Europe divided peacefully into separate parts, Norway and Sweden a hundred years ago, the Czech Republic and Slovakia more recently. The best solution for Ukraine would probably be a peaceful division into West and East. After that, East Ukraine might or might not be swallowed by Putin. West Ukraine might or might not join the European Union. The USA has no reason to be involved in either decision. It is not useful for us to tell Putin how to behave. We could give some help and encouragement to West Ukraine, but not tell West Ukraine how to behave. We should have learned from our experiences in Vietnam and Afghanistan and Iraq that we have as little power to control our friends as to control our enemies. I would like to be friends with Russia, but it takes two to make friends.

As this collection of letters is being gathered in memory of Professor Dyson, Russia has brutally invaded Ukraine again, making deliberate attacks on civilians, resulting in official allegations of war crimes and genocide. The outnumbered Ukrainians are fighting valiantly with admirable courage, tenacity, and unity. They fight for their families, their communities, their identity as a people, the very existence of their nation. They have "a cause clean to fight for." [*DU* Ch. 4] One recalls a line from *Weapons and Hope:* "Wars are fought by people, not by weapons…morale is in the end more important than equipment….[we should] learn to distrust any strategic theory which counts only weapons and discounts human courage and tenacity."[4] NATO is providing vast amounts of weapons but not troops. The Ukrainians are providing the troops, who display vast amounts of courage and spirit. Some cities in eastern Ukraine are home to citizens who speak Russian,[5] but the bulk of this post-Soviet and democratic nation leans towards the West. Putin's war is not a war demanded by the Russian people, so hopes for friendship between peoples may still be alive after the attacks stop, the criminals are put away, the dead are buried, and reparations have been made. We wish we could hear Professor Dyson's opinions today on the contemporary Russia-Ukraine conflict and prospects for their future relationship.

STS to FD:

4. Since science is evidence-based reasoning, do you think that some scientists deny the existence of God due to the lack of tangible evidence for God's existence?

Another way to word this question might be, do some scientists see *all* the dimensions of human experience through only the lens of science? We discuss science and religion issues near the end of the semester. Your letter to us about science and religion being two windows for looking at the world continues to be so very helpful. This topic never fails to hit close to home for many students here. In class we have discussed the importance and necessity of honest doubt, and how the existence of a Cosmic Mind cannot be proven in a way that will satisfy all reasonable observers. One cannot help but wonder, if God is real and has the attributes claimed by Christianity, why isn't He obvious to everyone? Sometimes the best we can do is to understand the questions; the answers might always remain elusive. Then we have to learn to live within the questions. In the absence of evidence that decides a question, one may choose what to believe, provided that one's choice is consistent with knowledge. Some of your comments in your speech "Science and Religion" that you delivered to the Conference of Catholic Bishops in 1986[6] have been helpful, and the closing chapter of *Disturbing the Universe*, with the scene of the three-month-old baby on the throne, whose smile swept the questions away, forms an appropriate benediction to our discussions on these matters. The majority of our students were raised in devout Christian homes (although I am glad to report that diversity among SNU's student population is increasing). For many of them, questioning their faith is a significant intellectual challenge, but a necessary one to go through if one is going to be honest.

Thank you, Professor Dyson, for your contributions to our class, for helping this professor teach it, and for being a friend to an entire generation of STS students. The STS class of the Spring 2014 semester wishes you joy and continued health, and extends our best wishes to Imme, to your children, and to all your wonderful grandchildren.

FD to STS:

4. Why do some scientists dogmatically deny the existence of God?

I make a strong distinction between agnostics, who doubt whether God exists, and atheists, who are sure that God does not exist. I agree with your discussion of science and religion. I consider militant atheists like Richard Dawkins to be just as misguided as militant fundamentalists who consign unbelievers to Hell. I was brought up as a church-going Christian by parents who were agnostics. A good scientist may be a religious believer or an agnostic, but should not be a militant atheist or a militant fundamentalist.

Just two days ago I flew home from a trip to Japan, a marvelously beautiful country full of wonderful people. I spent a day there with a group of students like you, talking about the same problems. The student who talked the best English was Tomone Watanabe, a nursing student who will graduate this year. By a happy coincidence, when I looked at my e-mail after returning home, I found a message from Tomone together with your message, asking similar questions. In spite of her excellent English and French, she has never traveled out of Japan. She belongs to the small minority of native Japanese Christians, and she takes her religion seriously. After she graduates, she will spend a year at the London Institute of Tropical Medicine. From there she will study public health problems in Africa and in Europe. I will not be surprised if she turns out to be another Mother Teresa. She has what it takes.

The answer to your last question is no. The majority of scientists are agnostics, because of the lack of tangible evidence for God's existence. Those who are militant atheists, dogmatically denying the existence of God, have a different motivation. They are driven by hatred for people whose feelings and ideas they do not understand.

That is enough for today. In conclusion, thanks again for your friendship and your questions. Imme was with me in Japan and enjoyed it as much as I did. Our big family is growing up fast and enjoying the American Spring.

Yours ever, Freeman

In the summer of 2014, Professor Dyson received a message from a translator, Nuria Reina, in Madrid who had translated *The Magic City* by Edith Nesbit [*DU* Ch. 1] into Spanish. When he replied to the translator, he generously copied us:

FD to translator:

July 13, 2014

Dear Nuria Reina,

Thank you for your friendly message. I am delighted to hear that you translated "The Magic City" and understood the deeper meaning of the story.

Now we have a new example of the same story. My grandchildren are playing with a beautiful new toy, a radio-controlled helicopter with four propellers that they can control accurately with two little switches. The toy is amazingly cheap (30 US dollars) including an electric charger to recharge the engines. It is light and rugged and safe for children to play with. It keeps them happily playing either outside or inside the house. The propellers are so light that a child can grab them with bare fingers and will not be hurt.

Unfortunately, somebody had the clever idea of using the same design of a radio-controlled helicopter to kill people. All you have to do is make the machine bigger and put an accurately-controlled missile on board, and you have a killer drone airplane. And now these killer drones are killing people in Pakistan and many other places. My grandchildren's beautiful new toy has become a permanent part of modern weaponry.

Thank you for your work as a translator. Translators are making a big contribution to the world community, helping us all to understand one another.

Yours sincerely,

Freeman Dyson

Not only is every technology weaponized—which says more about human nature than it says about technology—but every technology becomes more than just another tool. It can change the user's moral code. In can make evil increasingly anonymous. [*DU* p. 30]

STS to FD:

26 November 2014

Dear Professor Dyson,

On behalf of the Fall 2014 Science, Technology & Society class I bring you warm greetings. Once again we have journeyed with you through the landscapes and personalities and issues we find in *Disturbing the Universe*…

1. It has often been observed that we learn more from failure than we do from success. What have you learned from failure?

FD to STS:

3 December 2014

Dear Dwight and students,

I am back from our five-day trip to celebrate with daughter Mia and her family in Maine. The high point was a performance of "A Christmas Carol" in a local theater with twelve-year-old grandson Aidan on stage…

Here are answers to your questions. As usual, the questions are more interesting than the answers. The most important part of your education is to work out your own answers.

1. My biggest professional failure was in 1952. I was then a young professor with an army of graduate students working on a theory of the nuclear forces, the strong forces that hold the nuclei of atoms together. We used our theory to fit the experiments of Enrico Fermi, a famous physicist who measured the nuclear forces in

Chicago. This was a big deal. If Fermi accepted our theory, we would have solved the most important problem in nuclear physics.

We thought we had a good fit to Fermi's experiments, and so I took a Greyhound bus from Cornell to Chicago to show our results to the great man. Fermi received me politely but was not impressed by our results. He said "There are two ways of doing calculations in physics. One way, which I prefer, is to have a clear physical picture. The other way is to have a consistent mathematical formalism. You have neither." Fermi was of course right. He had the insight to see that our theory was no good. I sadly took the bus back to Cornell to tell the students that all our hard work was worthless.

I learned from this failure what my mother had told me long before. To enjoy any sport or any competitive occupation, whether it is football or science, the most important thing is to be a good loser. It is the good losers who make the enterprise enjoyable for everybody. This is especially true in science, which is an international game that everybody is free to play. By being good losers, my students and I were able to stay in the game and find useful things to do. Twenty years later, we could share the joy of the next generation of winners, when they finally solved the mystery of nuclear forces.

The aftermath of the 2020 US presidential election with the attempted coup of January 6, 2021, shows the essential importance of good losers to democracy.

STS to FD:

2. In "The Argument from Design" [*DU* Ch. 23] you discuss three roles for "mind" in the universe. The relation between mind and matter, brain and consciousness, has been a long-standing philosophical debate. Do the roles for mind in the universe depend on how one answers the mind-body problem?

FD to STS:

2. My answer to this question is no. I consider the role of mind in the universe to be a religious mystery, while the relation between mind and matter in a human brain is a scientific mystery. The two mysteries are both concerned with the nature of mind, but they are different. The mind-body problem might be solved using the tools of science, but the understanding of the mind-body problem would not give us understanding of the mind-universe problem. The mind-universe problem has nothing to do with the tools of science. To me it is important to keep science and religion separate. The idea that God might be explored by doing scientific experiments is absurd. The idea that the working of the human brain can be explored by scientific experiments is not absurd. That is the essential difference between religion and science.

STS to FD:

3. We know that you and your sister Alice were very close. We offer our condolences to you and to your family for her passing. What would you like for us to know and remember about Alice?

Fig. 18.4. Alice Dyson (1921–2013). Photo courtesy of Imme Dyson.

FD to STS:

3. You ask about my sister Alice. The main thing that you might find surprising about her is her religion. She grew up in the Church of England, which is in England the church of the upper classes, especially the church of old families who have been rich for several generations. In England the division of people into classes is sharper than in America. When World War II started, she became a professional social worker, taking care of patients who were not upper-class. As a result, she rebelled against her upbringing and became a Catholic. In England the Catholic church mixes the classes. It includes a few old upper-class families, with many more working-class families and especially Irish immigrants. The Catholic Church gave her much better contact with her patients, and especially with the Irish priests who helped her to take care of them. As a result of her change in religion, she moved away from the snobbish narrow circle of friends that we grew up with. She moved into a different crowd that she found far more congenial, including policeman and criminals and their girl-friends and children, people who had difficult problems and really needed help.

My most vivid memories of Alice come from days when I drove with her in her car doing house-calls. She specialized in unmarried mums, who need two kinds of help, for the mother and for the baby. She liked to take me with her on house-calls, so that she could talk seriously with the mother while I played with the baby. These were happy experiences for both of us. My other vivid memory is of the Catholic church where I would go to Sunday Mass with Alice and pretend to be a Catholic. At the memorial Mass after she died, about two hundred of her friends and family came to the service.

STS to FD:

4. Are we becoming too dependent on automation? A reference we have used on this topic is the new book by Nicholas Carr, *The Glass Cage: Automation and Us.*[7] Carr begins by describing a January 2013 Federal Aviation Administration Safety Alert for Operators.... When the automation fails, the pilots suddenly find that they have forgotten how to fly the airplane manually. Now we see the same trend being relentlessly foisted on all of us, towards turning our skills and responsibilities over to software, such as autonomous cars, networked appliances in households,

and so on. In addition to the de-skilling, our relationship with our machines gets drastically changed. Carr writes, "Pilots have always defined themselves by their relationship to their craft. Wilbur Wright, in a 1900 letter…said of the pilot's role, 'What is chiefly needed is skill rather than machinery.'" Carr continues, "As we begin to live our lives inside of glass cockpits, we seemed fated to discover what pilots already know: a glass cockpit can also be a glass cage." This reminds us of "The Magic City" [*DU* Ch. 1]: once we wish for machinery we are stuck with it. But it seems now that we are stuck with it even if we do not wish for it; it is foisted upon us even when we do not ask for it. What are your thoughts on automation and what it does to our abilities to be self-reliant?

FD to STS:

4. I have not read the Nicholas Carr book, but I agree with your remarks about it. I recommend that you read another book, "The Human Use of Human Beings," published by Norbert Wiener in 1950, an amazingly far-sighted view of the problems that would arise from automation. Wiener was writing before electronic computers existed, but he was expert in mechanical control systems, and he foresaw the dangers that would come from the combination of electronics and control systems. His most famous prediction was made on page 189 of the book.[8] *"The automatic machine is the precise economic equivalent of slave labor. Any labor which competes with slave labor must accept the economic conditions of slave labor. It is perfectly clear that this will produce an unemployment situation, in comparison with which the present recession and even the depression of the thirties will seem a pleasant joke." This is a different problem from the loss of skills and loss of self-reliance discussed by Carr, but the loss of skilled blue-collar and white-collar jobs goes together with a loss of self-reliance. This is a permanent loss, and will make any lasting economic recovery impossible so long as our politicians believe that market economics will solve our problems.*

In my opinion, any lasting recovery will require that we abandon market economics applied to human labor. We must take human labor out of the market-place. Let workers be paid according to their need, even when machines can do the job more cheaply. This makes sense, even though Karl Marx said it. Other parts of Marx, such as the dictatorship of the proletariat and the public ownership of the means of production, do not make sense. Unfortunately the disasters caused by Marxist nonsense have discredited the Marxist ideas that make sense.

Fortunately, there are two human enterprises that will always depend on human skills and keep highly skilled people productively employed. The two skill-driven enterprises are art and science. In the future, as machines take over more and more of the routine work required by a modern society, more and more of our young people will be artists and scientists. In the future, as in the past, the great civilizations will be those that give art and science sustained and generous support.

STS to FD:

5. In November 2007 our STS class asked you what you thought about the US stationing weapons in space. You pointed out that with communication and surveillance satellites in orbit for military applications we already have weapons in space. We would like to narrow the definition of weapons in space and ask the question again if you don't mind. Let us define weapons in space as systems envisioned by Air Force generals[9] following the 2001 Rumsfeld Commission recommendation which stated that "The US government should vigorously pursue the capabilities called for in the National Space Policy to ensure that the president will have the option to deploy weapons in space." The following year President George W. Bush withdrew the US from the Anti-Ballistic Missile Treaty, sweeping aside legal impediments to the US stationing offensive weapons in low-earth orbit.

The meaning of "weapons in space" we propose comes from statements by US Air Force leaders made after the

Rumsfeld Commission's report. The *NY Times* (18 May 2005) quoted a couple of them:

*Pete Teets, then Acting Secretary Air Force: "We haven't reached the point of strafing and bombing from space. Nonetheless, we are thinking about those possibilities."

*General Lance Lord, commander of the Air Force Space Command: "...we must establish and maintain space superiority. Simply put, it's the American way of fighting." The *Times* article went on to describe space superiority as "freedom to attack as well as freedom from attack." General Lord continued, "Space superiority is not our birthright, but it is our destiny."

On 11 January 2007 the Chinese shot down one of their weather satellites, presumably to demonstrate that if the US wants to launch an arms race in space, they can play too.

We do not see how such policies can be consistent with "The Ethics of Defense." [*DU* Ch. 13] If we place in orbit the kinds of weapons that General Lord wants, we think that will guarantee other nations following suit. Then we will be back into another multi-trillion-dollar arms race similar to what we saw during the Cold War. What are your thoughts on this issue? To ask this question another way, what would you say to General Lord and his like-minded colleagues?

FD to STS:

5. I had not heard the remarks that you quote from General Lord. They are an extreme expression of the Air Force mentality that goes back almost a hundred years, to the slogan "Victory through Air Power" that drove the air forces (then divided between the Army and Navy) in World War II. The idea is that we can win wars and enforce our political objectives by maintaining air superiority and killing our enemies from the sky. The idea is wrong and dangerous, now more than ever. It is unpractical because weapons in space are highly vulnerable and highly visible. Any attempt to achieve space superiority over China would be successfully resisted. The result would be retreat and withdrawal if we were lucky, a major war if we are unlucky.

To answer your question, the policies advocated by General Lord are totally inconsistent with "The Ethics of Defense." If I had the opportunity to talk to him, I would tell him that he is wrong militarily, wrong politically, and wrong morally. Wrong militarily because his policies give our enemies easy targets to shoot at. Wrong politically because bombing attacks and drone killings strengthen our enemies and weaken our influence. Wrong morally because his policies blur the distinction between war and peace, keeping the world in a permanent state of tension and unrest.

STS to FD:

6. What was your role (perhaps as a member of Jason?) in convincing President George H.W. Bush to stop nuclear weapons testing, which led to the present policy of Stockpile Stewardship?[10]

FD to STS:

6. I do not claim to have had any influence on the decision to stop testing. So far as I remember, the decision was taken by President Bush and three other people, the directors of the three weapons laboratories, Los Alamos, Livermore, and Sandia. The three directors had to promise to Bush that they could continue to provide reliable weapons without testing. Like other bureaucrats, the directors were mainly concerned with maintaining the budgets for their organizations. So they took this opportunity to do a deal with Bush. The agreed to promise to provide reliable weapons, and they demanded that Bush promise to provide reliable funding. To provide reliable funding for the weapons labs after they stopped testing, Bush agreed to establish the Stockpile Stewardship program. In my opinion, Stockpile Stewardship is not really a program. It is a political device to make sure that the weapons laboratories get a generous and reliable level of funding. It is not needed for technical reasons. The bombs are reliable and durable enough without any stewardship. Stockpile Stewardship is needed for political reasons, to make sure that the lab directors do not demand a resumption of testing. However, Stockpile Stewardship was officially approved by Jason, and many of my friends sincerely believe that it is essential to our security. It is probably worth paying that price to get the testing stopped.

STS to FD:

7. Your description of "absolute silence" at Jackass Flat, Nevada, [*DU* p. 128] offers an opening for discussing questions about silence (both audio and visual), such as: Why is silence so difficult to find in our society; When we do find it, why are we so quick to drown it out; Why do we not push back hard against the in-your-face noise to which we are continually bombarded; Why is our eagerness to embrace the latest technology not accompanied by reflection over what it displaces—and so on. Professor Dyson, where do you go, and how often do you go there, to be alone with your own thoughts, to find a moment of silence?

FD to STS:

7. I am lucky to have an office at the Institute for Advanced Study where I work. Although I am retired, they let me keep the office as long as I can make use of it. I usually keep the door open so that the young Institute members are not afraid to walk in. But whenever I need to have silence, I can shut the door. I usually shut the door only when a visitor is with me and wants to talk privately. But I can also shut it when I am alone. I remember long ago reading the book, "A Room of One's Own," by Virginia Woolf. That book was written a hundred years ago when families were bigger and fewer people had rooms of their own. It encouraged a whole generation of women to claim the right to silence. I am lucky to have silence available whenever I want it. I do not need it to be as quiet as Jackass Flat.

I find it horrible to be at a party with loud music, often so loud that it is impossible to have a serious conversation. The habit of playing loud music at parties is doing permanent damage to the hearing of young people. It is also depriving them of the opportunity to have serious conversations. Since I always like to look on the bright side of things, I rejoice to see that the railroad on the regular route through New Jersey has "Quiet Cars," where passengers may not use cell-phones or carry on conversations. These cars are not totally silent, but they allow passengers to read or work in peace. They are a good step in the right direction.

STS to FD:

8. The STS class of last April asked you what social reforms will be necessary if nuclear energy turns out to be our only long-term energy resource that can meet the needs of national and global economies. That question grew out of our discussions of environmental sustainability, including our dependence on non-renewable fossil fuels. Since our economies are dependent on a resource that will no longer be available within another human lifetime or two, we wondered last April if such dependence qualifies as "insanity," given the definition of insanity that you offered in "The Greening of the Galaxy:" [*DU* Ch. 21] "Sanity is, in its essence, nothing more than the ability to live in harmony with nature's laws." [*DU* p. 237]

Your reply was very instructive because you put people first...You went on to describe how fossil fuels are enabling China and India to become rich, and in the meantime we should develop solar energy so that when fossil fuels are depleted we will have sustainable, adequate energy for all. You said "Political actions to make fossil fuels more expensive make inequality worse." Point well taken.

We appreciate that you put people first. We understand too that preserving the environment means that today's immediate needs must already be met, so that we can afford to set some parts of nature aside. Henry David Thoreau expressed it well when he wrote in *Walden,* "A man is rich in proportion to the number of things which he can afford to leave alone."[11] Thus the problems of poverty and environment form a strongly coupled system.

At last we come to this question from the Fall 2014 class: We would like to enlarge the scope of last spring's question about environment and insanity. We see ongoing destruction of forests, continuing pollution, over-fishing in the oceans, loss of wildlife habitat and biological diversity, and so on. At what point will our consumptive, short-sighted lifestyle tip over into insanity as you have defined it?

FD to STS:

8. I answer this question the same way as I answered your question in April. I think you are wrong in seeing our life-style as only destructive and short-sighted. In the real world, we are doing a lot of destruction and a lot of preservation. The media give us a false impression by giving us only the bad news and not the good news. One of the advantages of being ninety years old is that I can see the good news more clearly.

I happen to live in New Jersey, a state that is heavily populated and heavily industrialized. It is proud to give itself the name "The Garden State." Fifty years ago, visitors to New Jersey considered the name to be a joke. On the main highway driving South from New York, there is an enormous oil refinery. Fifty years ago, the whole region near the refinery stank of sulfurous fumes. Visitors had to shut their car windows and drive through as fast as possible. Now the refinery is still there but the stink is gone. The Garden State is no longer a joke. The state has learned how to clean up a bad mess. All that it takes is time and money and political will.

The forests in New Jersey are also growing. A hundred years ago, the natural forests had been destroyed, mostly to grow hay for the horses in nineteenth-century cities. Now the hayfields are mostly growing back to forests, and the animals and birds that live in forests are returning. In the last ten years, a substantial population of bears has returned to the state. At the same time, a substantial population of humans has returned to small-scale farming as a hobby. Friends of ours near to Princeton are raising cows on grass, producing meat that tastes better than meat from animals raised on feed-lots in Iowa. Other friends are campaigning to preserve wetlands and bird sanctuaries. I do not see any signs of our population tipping over into insanity.

Of course, rich people do better than poor people in caring for the environment. That is why the worst destruction of nature is happening in the poorest countries. The poor people are not insane.

> *They only lack the time and money that are needed to save forests and fish.*
>
> *Thank you once more for a good set of questions...*
>
> *Happy Christmas to you and the class!*
>
> *Yours ever, Freeman*

The question we asked—twice—about whether our society is insane has been very stimulating to classroom discussion and personal thought. Once again we see the importance of consulting our tribe's elders, of "walking with Grandfather."[12] In reflecting over this last answer to our questions, the *Dear Professor Dyson* account summarized it this way:[13]

> One might argue that our society *would* be insane if we never changed course, cutting and slashing with no regard to consequences. But while some sectors of society seem to do that, other sectors push back. For every Monsanto and Wal-Mart there is also a Greenpeace and a Sierra Club... American society, in particular, seems to operate in the Pearl Harbor mode of response: Despite warnings, very little is done until the crisis strikes, but when it strikes we respond with everything we have. While not a reassuring strategy, in the past it has been effective. We can only hope that our grandchildren, and their grandchildren, will be able to enjoy starry skies and pristine landscapes that we have been privileged to experience. As Professor Dyson observed, our sense of brotherhood with all mankind is an anchor "essential to our sanity." [*DU*, p. 169]

If our grandchildren cannot enjoy starry skies or pristine landscapes, then they will be the victims of colossal theft. We remember that we are not owners, but stewards. Ultimately, we do not *own* anything. We merely *borrow* it from our children and our grandchildren.

20 Social Justice as Necessary for a Healthy Society

In the universe there floated an island; it looked like earth. On the island lived two sorts of people, the rich and the poor… One school was designed for the rich, who were to learn the art of giving. ….The other school was designed for the poor, who were to learn the art of taking….The rich did not learn giving. They learned what was designed for the poor, cheerfulness and pride; and it was only the better ones among them who felt ashamed of their cheerfulness and did not allow their pride to degenerate into arrogance.

The poor did not learn taking. They learned what was designed for the rich, sadness and humility; and it was only the better ones among them whose sadness did not turn into hatred…. Now as before the rich and the poor attended the wrong schools and did not know it. The Book of Joy has been lost.

—Fritz Mauthner[1]

STS to FD:

22 April 2015

Dear Professor Dyson,

On behalf of the Spring 2015 class, we bring you greetings, with good wishes to you and all your family…

The students in this semester's class discussed several candidate questions last night, and have narrowed the field to five…

1. Your mother ran a family-planning clinic in Winchester, before such clinics were as abundant as they are today. In this role how was she, her clinic, and her clients received by the local community? As part of our discussions about genetics, last night we also talked about the abortion debate that has gone on so long in society. Picketing abortion clinics is not unheard of in this region. We are not sure if your mother's clinic was involved in abortion services, but discussions about abortion led to us wanting to know more about her work with family planning. Your mother was a

pioneer in this field. Was Alice's work with single mums related to your mother's work with families?

FD to STS:

1 May 2015

Dear Dwight and students,

This has been a busy week. The main event was a big celebration for our friend Oliver Sacks in New York.[2] *You may have read some of his books. He learned a few weeks ago that he has a spreading melanoma that will kill him in a few months. So he threw a big party for his friends to celebrate his life. It was a wonderful and joyful gathering. Oliver was in great spirits, and we were too.*

Now I must try to answer the student's five questions…

1. I know very little of the details of my mother's birth-control clinic. The only episode that I remember vividly is when I was six or seven years old, (Fig. 20.1) *I picked up some paper from her desk to write my school home-work.*

Fig. 20.1. Freeman Dyson, at the age for doing homework. Photo courtesy of the Dyson family.

On the back of the paper was printed, "Winchester Birth-Control Clinic." As I was walking out of the house to school with my home-work, my mother happened to see the back of the paper, and she told me fiercely that I must not use that paper. Being involved with birth-control in those days was not socially acceptable.

I do not know whether the clinic was arranging abortions. I would guess that it was not. My mother was a lawyer and not a physician. I would guess that the clinic was mostly concerned with educating young girls about the facts of life, and arranging adoptions of unwanted babies. I remember my mother telling us about one of her clients who had a baby with bright red hair. My mother asked her whether the father also had red hair. The mother replied, "I couldn't rightly tell you, Ma'am, he had his hat on." My mother was an upper-class English lady. She saw her clients as children who needed to be helped and guided, butnot as equals.

My sister became a medical social worker during World War II when every young person had to do some kind of national service. I do not know whether my mother had anything to do with her choice of service. My sister was a trained professional. She loved the work and stayed with it all her life. She worked in a hospital and dealt with all kinds of patients with all kinds of problems, not only unmarried mums.

STS to FD:

2. What professions would you most recommend to twenty-somethings starting their careers today? Would you encourage them to lean towards technology-dependent professions, or to careers that require the maintenance of skills that cannot be done by automation?

FD to STS:

2. I would advise young people to be prepared to switch from one job to another every few years. Concentrate on acquiring skills that remain useful in different jobs, such as writing clear English, understanding simple mathematics, managing financial accounts, organizing computer networks, running committees. There are of course exceptions. A few professions, for example, medical doctor, minister of a church, university professor, give a good chance of finding a secure and permanent job. But most jobs are not secure and not permanent. The disappearance of jobs is not predictable. It is not only caused by automation or by new technology. So the important thing is to be prepared to switch. I like to recommend my daughter Esther as a role model. She began working as a journalist for Forbes, then switched to working as a technical analyst for the banking firm Oppenheimers, then switched to publishing her own business news-magazine, then switched to running her own venture-capital business, and now recently switched to running her own private health-care foundation. Her motto which is printed on every E-mail says "Always make new mistakes." The thing is to avoid making the same mistake over and over again.

STS to FD:

3. What should the USA, NATO, and other concerned countries do about ISIS? You wrote in a previous letter that terrorism is ultimately a problem about people's hearts and minds, and thus military action alone will not solve it. If the atrocities and excessive certitude of ISIS could not pull diverse nations from the US to Iran together in common cause, it's hard to know what could.

FD to STS:

3. I do not claim to know anything about ISIS or how to deal with ISIS. My guess is that the present policy of trying to suppress ISIS by military force will not succeed. If I were in charge, which fortunately I am not, I would leave ISIS alone for a few years and see what happens. I would also make friends with Iran, which seems to me to be the most stable and civilized country in the Middle East. I would also not get excited if Iran decides to build some nuclear weapons. It was stupid for us to get so deeply involved in the Middle East in the first place.

The next question was an offshoot of a letter our class wrote to one of Oklahoma's US Senators[3] regarding America's nuclear policies. Our letter asked the Senator, who chaired the Senate's Armed Services Committee, why the US keeps *thousands* of nuclear weapons. Given that it took only two nuclear bombs to end World War II, wouldn't 10 or 100 be more than enough? At whom would we shoot the other thousands? Why not unilaterally reduce our stockpile by 10% or 25% and see if other nations respond similarly, and continue in this way until the number of nuclear weapons in the world's arsenals approach zero? Before sending our letter to the Senator we asked Professor Dyson to review it.

STS to FD:

> 4. Attached to this message is a draft of a letter to our US Senator…The class made some suggestions for the letter (e.g., one student suggested "The USA considers itself to be a leader, so reducing our nuclear arsenal unilaterally offers an opportunity to show leadership")… If you have a moment to look it over and make suggestions, corrections, additions or deletions, the class and I would be very grateful… You have much experience in conversations on these matters with powerful people who know more about them that the general public….

Professor Dyson clarified our understanding of the types of nuclear weapons that have been decommissioned in recent decades, and brought us up to date on the composition of the contemporary US arsenal:

FD to STS:

> *4. The fact is that ten-megaton hydrogen bombs were taken out of the stockpile a long time ago. One-megaton bombs were taken out more recently. Today the biggest bombs in the stockpile are less than half a megaton. The Air Force generals and Navy admirals understood a long time ago that small bombs are more useful than big bombs, and so the bombs in the stockpile have been getting smaller as time goes on. At the time of the big debate about hydrogen bombs when [Robert] Oppenheimer spoke, everyone was thinking of the hydrogen bomb as a ten-megaton monster. That is no longer true. I like to say, if you look at the stockpile today, it is almost the same as it would have been if the hydrogen bomb had never been invented. In the end, the big debate about the hydrogen bomb turned out to be unimportant. If we had only fission, the bombs would look almost the same.*

Of course, even if you do not have hydrogen bombs, half a megaton kills a lot of people, and there are plenty of reasons why it would be a good idea to get rid of it...

With all my good wishes and thanks to all of you, yours ever, Freeman Dyson

We eventually received a response from our Senator. He (or perhaps a staff member) sent what appeared to be standard boilerplate, merely telling us how America "must remain strong." Our good Senator was tone-deaf to our suggestion that the US make the first move, even incremental steps, towards rendering nuclear weapons obsolete.

STS to FD:

5. If you could "get away from it all" anywhere in the world, where would you go and what would you do? We know that you and Imme enjoy traveling, and have been to practically every country on the planet, including the Galápagos Islands. This question is similar to a recent one we asked you about where you go to find a moment of silence. But the difference between the two questions, we think, is asking what kinds of environments on this planet would most resonate with who you are deep down (we know you would also like to go to Saturn [*DU* Ch. 10] and would not mind visiting an asteroid).

Thank you Professor Dyson for your consideration of our questions. Do not feel obliged to answer all of them. We do not want to take your time for granted.

Warm regards, the STS class

FD to STS:

5. I cannot give a useful answer to this question because I would hate to "get away from it all." I need friends and I need family and I need the community in which I lived for sixty years. Humans are social animals, and most of us like to be constantly interacting with friends and family and community. I enjoy visiting exotic places like the Galápagos, but even there it is not the solitude that I enjoy. I enjoyed the Galápagos because I lost my suitcase and some young girls on the boat lent me their colorful blouses, so my wife and I joined the group of young people and became friends right away. Losing the suitcase turned out to be lucky. We even found it again when we flew back to Quito. I can only answer the question by saying, I would love to get away from it all for five minutes, but not longer.

By the way, I just published a new book with the title "Dreams of Earth and Sky,"[4] *a collection of book reviews published by the New York Review of Books. Yesterday the local book-store in Princeton had an evening session, half an hour of me reading aloud from the book, then half an hour of question and answer discussion, and finally a book-signing with me signing books for people who bought them. I was amazed to see how many books we sold. A big crowd came and they were all my friends. That is why I like to stay here and not get away from it all.*

STS to FD:

6. If you were granted the privilege of being God for a day, what would you do during that day? The students who proposed this question said to assume the properties usually attributed to God, such as omnipotence, omnipresence, and so on.

FD to STS:

6. If I were God, I would do what God has been doing for the last few centuries at least, watching the show and not actively interfering. Any quick or hasty actions would probably do more harm than good. God has given us the freedom to make mistakes and learn from our mistakes. I would continue with the same wise policy.

With all my good wishes and thanks to all of you, yours ever, Freeman Dyson

STS to FD:

5 April 2016

Dear Professor Dyson,

The Spring 2016 STS class brings greetings and good wishes to you, to Imme, and to your family....

...The Spring 2016 STS class has assembled a few candidate questions.... We never want to get in the way of "family first...."

1. In our discussions about science and religion, faith and doubt, some of us have found that our early opinions and beliefs are challenged as we acquire more life experience and wider perspectives. In your life have you ever believed something deeply, but changed your mind because of new evidence or insight? If so, was the transition difficult?

FD to STS:

16 April 2016

Dear Dwight and students,

Thank you for your message and your questions. Thank you also for the "Dear Professor Dyson" book which I am enjoying enormously. It reminds me of many good discussions with your predecessors. Here are the answers to the latest batch of questions.

1. Since you have read "Disturbing the Universe," you already know about the two strong beliefs that dominated my thinking as a teen-ager, the religion of "Cosmic Unity" [DU Ch. 2] *that solved the problem of evil, and the political belief in pacifism that solved the problem of war and peace. I explain in the book how those two beliefs were abandoned. Cosmic Unity was gradually abandoned because none of my friends wanted to hear about it, and I could see that it did not make sense to start a new religion with only one believer. Pacifism was abandoned less gradually when France was defeated by Germany in 1940, and I could see that the genuine pacifists in France could not be distinguished from the collaborators and opportunists. It quickly became clear that pacifism in England would be indistinguishable from collaboration with Hitler. So I reluctantly joined the majority who were fighting Hitler with unexpected good humor and success.*

There was a third strong belief that lasted longer and is not discussed in detail in "Disturbing the Universe," namely socialism. When Churchill became Prime Minister in 1940, he was head of a coalition government, with the two main parties, Conservative and Labor, working together. The Labor party was socialist. Churchill made a deal with the socialists. He would run the war and they would run the country. This arrangement worked amazingly well. All through the war, the domestic economy was socialist. Food and clothes were rationed and cheap. Rich and poor got the same rations and shared the hardships equally. Money was unimportant because there was not much to buy. Class distinctions were also unimportant because there were no servants and no luxuries. The war years in England proved that a socialist society is possible and can bring social justice. So I remained a believer in socialism, and I am still a socialist today. I still believe in social justice as necessary for a healthy society. But I have learned from living in a capitalist country that capitalism also has its good side. My daughter Esther is a venture capitalist and has put her wealth to good use, helping small companies to get started and to survive. She is now engaged in a venture to improve public health by a bottom-up organization based in local communities. Her free style of operation would not be possible in a socialist state-controlled society. So I have now come to believe in a compromise between social justice and economic freedom. I would like to live in a country that combines socialist ethics with capitalist enterprise, attempting to give us the advantages of both.

STS to FD:

2. We know that you have been a controversial figure regarding the consequences of climate change. But every public discussion needs someone who says "Not so fast, let's think carefully about this." As with all changes, some consequences will be bad (e.g., flooding low-lying areas), others will be good (e.g., Greenland's economy moving more towards tourism and relying less on fishing, and thus not losing a third of their young men, as you described in an earlier letter). You have made the point well that any major

actions should be based on evidence from the real world, and not from computer models only. You have also made the important point in previous letters that a warmer climate may be the price for bringing much of humanity out of poverty. With all that background, are you happy with the outcome of last December's Climate Change Conference in Paris?[5]

FD to STS:

2. *I have not much to say about the Paris Climate Change Conference. I am happy with the outcome, because it does not impose any strict regulations on anybody. It leaves each country free to choose its own climate policy. There are a lot of fine words about reduced burning of fossil fuels, but no binding promises. I am glad that the Paris conference made a silent decision not to revive the Kyoto treaty,*[6] *which had tried unsuccessfully to reduce fossil-fuel burning by regulation.*

In my opinion, the countries that might have a major effect on climate are India and China. They are the countries with large and rapidly growing economies, depending on coal for growth. India and China have a clear choice. Either they continue to get richer by burning large quantities of coal, or they reduce their coal-burning and stay poor. I hope they will choose to get richer. I believe that the rise of India and China to become rich countries is the greatest historic event of the twenty-first century. But the choice belongs to them and not to us. It is not our business to tell them how to behave. Whatever we may decide to do with our own renewable and non-renewable sources of energy will not make much difference.

STS to FD:

3. You have known so many interesting people in your life. You have shared with us in *Disturbing the Universe,* and in your letters, stories about your parents, your experiences with Hans Bethe and Richard Feynman, Robert Oppenheimer and Edward Teller. You knew John von Neumann and Jacob Bronowski. You are good friends with Steven Weinberg; at Cambridge you took courses from Paul

Dirac and your office was down the hall from Ludwig Wittgenstein. Your children have all distinguished themselves in various professions. Imme has been an inspiration to us too (e.g., her discipline in running ten miles a day and competing in marathons). The stories you could tell about the people you have known, some famous, some not, but all interesting, could fill several books. Do you have an unpublished anecdote about someone that you would feel at liberty to share, that would offer an interesting life lesson or a good story?

FD to STS:

3. For an unpublished anecdote I choose Taro Asano, a brilliant young Japanese scientist who solved a famous problem in the theory of magnetism.[7] *After he did his great work in Japan, I invited him to come for a year to the Institute for Advanced Study in Princeton. He came with his young wife Sachiko and worked as a member of our Institute. Since he had his wife to take care of him, I assumed he was OK and neglected him for some months. During that time he tried hard to find another problem that he could solve. He failed to repeat his earlier success. He became seriously depressed and miserable. But I was not watching out for him.*

Finally, one afternoon, Sachiko came running to our house, telling us that Taro had grabbed their car-keys and was driving their car at a crazy speed. We started to run after him, but then we heard a terrible crash. He had driven his car at high speed into a head-on collision with another car coming in the opposite direction. The other car carried a family of six people. They were a Chinese family, also belonging to the Institute. Luckily, Taro's car was light and the Chinese car was bigger and heavier. Taro was dead, but the Chinese were all alive with only a few broken bones. I apologized to the Chinese mother, but she was very cheerful. She said, "Oh, I don't blame you for this. The Japanese are all crazy." Of course I did blame myself. I was responsible for bringing Taro to a strange country at a difficult time in his life, and I did not take care of him. It was my fault that this brilliant young life was destroyed. The guilt for Taro's death will always remain with me.[8]

Luckily the story has at least a partly happy ending. Taro was cremated according to the Buddhist ritual, with Sachiko and me and my wife each picking out pieces of bone from his ashes. After that, Sachiko carried the box with the ashes everywhere she went. A week later, I flew with Sachiko and the box of ashes to bring her back to her family in Japan. We arrived at the Narita airport and found her family there waiting for us. She ran back to her family without a word of good-bye. According to Japanese custom, she mourned for Taro only for 49 days. After 49 days, she went back to her maiden name and was free to remarry. When last I heard, she was in good health and happily remarried.

STS to FD:

4. The STS class of last spring asked you what the US, NATO, and other concerned countries should do about ISIS. You said that, if it were up to you, we should ignore ISIS. After the horrific mass murders in Paris and Brussels recently, politicians running for President have been breathing fire and brimstone about ISIS. One (Donald Trump) said he would "bomb the hell" out of them, another (Jeb Bush) said he would use this event as another opportunity to build up the US military even more. This kind of inflammatory rhetoric is probably just what ISIS wants, since it will drive more radicalized young people into their arms as they anticipate the great Armageddon-like confrontation espoused by their ideology. In a letter to us written soon after 9/11, you pointed out that terrorism is not a military problem, and military responses only make it worse. Rather, you said, rightly we believe, that terrorism is a problem of people's hearts and minds. Now to our question: Given the regime still in power in Syria, given the refugee crisis, and given recent events, is it too late to ignore ISIS even if our leaders wanted to?

FD to STS:

4. I do not know much about ISIS and do not claim to have a solution to that problem. Now ISIS has spread from its base in Syria and Iraq to France and Belgium, and it looks much more frightening to Americans. But I still think it should be handled as a local problem. It cannot be defeated by American armies and American drones fighting in foreign countries. By invading its home territory, we only make it stronger. It must be fought by each country defending its own territory. Defense in different countries will require different methods and will bring different outcomes. We cannot compel different countries to play the game by our rules, as we should have learned from our failures in Vietnam and Iraq and Afghanistan. We should defend our own people as best we can, and leave to other countries the responsibility for defending theirs. I am not saying that we should ignore ISIS, but we should not make it stronger by exaggerating its importance.

STS to FD:

5. The 2016 US presidential election cycle has been—shall we say—interesting—at the least. There is much blustering about "making America great again" with emotional appeals to nationalism and fear, hostility towards certain ethnic groups, and looking to the military to solve our problems. Some of the rhetoric and posturing reminds us of what we have read about German politics in the early 1930s.[9] In this campaign season there are also attempts to draw attention to the growing gap between rich and poor, but that does not make the evening news as readily as a violent fistfight at a rally. What do you make of the candidates and their followers in this election cycle, and what does this suggest about the future of this large and seemingly fracturing country?

Thank you, once again, Professor Dyson, for being a member of our class, and for your contributions to teaching it through *Disturbing the Universe*, your other writings and speeches, and above all your wisdom, forward-looking spirit of optimism, and appreciation of reality.

> Please give our warm regards to Imme... We appreciate receiving the January 2016 Dyson Family Chronicle. Best wishes too for the next chapter in your life when you will be dividing your time between Princeton and San Diego.
>
> Warm regards to you and the entire Dyson family, the Spring 2016 STS class

In the Dyson family's New Year's letter for 2016, Professor Dyson mentioned that, while he and Imme kept the family home in Princeton, they would divide their time between Princeton and San Diego, the latter location near one of their daughters. Of the San Diego apartment, they said it was "large enough for two but not too large for one."[10]

FD to STS:

> *5. I have nothing new or illuminating to say about the US election circus. I take comfort from the US Constitution. The founding fathers wrote the Constitution wisely so that the government can be run by crooks without doing too much harm. This is different from the German government in the 1930s which was designed to be run by gentlemen. It is true that Donald Trump talks and behaves in many ways like Hitler. The difference is that Hitler as Chancellor of Germany had the power to change the laws, put his opponents in jail, and make himself dictator, while Trump as US President would not have such power. President Trump would have to deal with a recalcitrant Congress and a conservative Supreme Court, just like President Obama. President Trump might do a lot of stupid things, but he could not put Hillary Clinton and Bernie Sanders in jail, and he could not spend money without the consent of Congress.*
>
> *I am delighted with the fact that Bernie Sanders is telling the truth about the gross inequality of American society. He will probably fail to win the Democratic nomination, but he has succeeded in getting the country to pay attention to his message. He has made it much more likely that some younger populist politicians will win future elections. Then there will be a good chance that the US will finally deal effectively with the social*

problems of inequality and greed. I do not expect that I shall still be around when that happens.

Thanks to the students for a good list of questions.

Yours ever, Freeman Dyson

21 The Serenity of Old Age

...And there is the silence of age,
Too full of wisdom for the tongue to utter it
In words intelligible to those
Who have not lived the great range of life.
—Edgar Lee Masters[1]

STS to FD:

15 December 2017

Dear Professor Dyson,

Congratulations on your 94th birthday! On behalf of the STS class, I bring good wishes to you and your family... We are sending you a pair of Christmas cards signed by the students, which also contain birthday greetings for you.

As we usually do, we gathered a few questions, should you have a moment to answer some of them. The most important questions we could ask are "How are you doing?" and "Are all members of your family well?" and "Do you have any more grandchildren?"...

1. You have seen so much innovation and technological change in your life. What piece of technology has created the most good for society?

FD to STS:

18 December 2017

Dear Dwight and students,

....First I answer the question about family news. I am now enjoying warm sunshine in California while you are still shivering in the snow. I met three of the grandchildren yesterday at their home in San Diego. The oldest is already a student in Berkeley, the youngest an eighth-grader in San Diego. It is a long time since we had babies in the family. We are lucky to be in San Diego, far away from the wild-fires that are raging in Santa Barbara. Next week we

shall visit another lot of grandchildren in Oregon. All of them seem to be doing well.

Now come the answers to your three numbered questions.

1. What piece of technology has done the most good for society in my lifetime? There are many important new technologies that have done good, for example the polio vaccine that saved many lives, and the cell-phone that keeps us in touch with friends and family around the world. But I give first place to the contraceptive pill that has given women all over the world better control of their own lives. For example, in my lifetime the size of the average family in Mexico has decreased from seven children to two and a half. Similar big decreases of family size have happened in many other countries. This makes the difference between a rapidly exploding and a stable population. Thanks to the pill, the population of the whole world is now growing slowly enough so that food-production can keep pace with it.

STS to FD:

2. Given the inescapable abundance of social media, do you think that the interpersonal social skills of future generations will be lessened due to these technologies?

FD to STS:

2. Will the spread of social media cause a loss of interpersonal skills in future generations? This question is impossible to answer, because new media cause both gains and losses in every generation. Looking at the past, humans began with a purely oral culture, reciting stories and poems around the cave-fire. They had far better verbal memory than we do.[2] *Then came the culture of writing and reading, with a loss of verbal memory but a big gain of long-range communication and a big gain in the education of children. Then came the telephone, with a big gain of speed of communication and a big loss of letter-writing skills. Now comes your generation growing up with Facebook and Twitter, giving you a gain of superficial outreach and a loss of deeper personal relationships. The social medium of the next century will perhaps be radiotelepathy, the direct communication of thoughts and feelings from one brain to another by means of microwave transmitters and receivers embedded in our neurons. Radiotelepathy would bring us a big gain of intimate personal friendship and understanding, paid for with losses of privacy and losses of verbal skills. We cannot yet imagine future media beyond the next century, but we can be sure that there will always be gains together with losses.*

STS to FD:

3. Are ideas created, or are they discovered?
Thank you, Professor Dyson.... We wish you and Imme, and your children and grandchildren, a joyous Christmas season.
Warm regards, STS class

FD to STS:

3. Are ideas invented, or are they discovered? This question is simple to answer. The answer is both. Examples of invented ideas are music, poetry, architecture, stories and theories. Examples of discovered ideas are numbers, geometrical patterns, ecological patterns, history and cosmology. Science is a prime example of the intimate mixture of invention and discovery. We discover Nature's ideas and invent our own. The central mystery that makes science possible is the agreement between Nature's ideas and ours. Science is the bunch of tools that we use to hear Nature speaking.

That is all I have to say. Thank you for your thoughtful questions. I wish you all a Happy New Year at SNU. Yours ever, Freeman Dyson

On February 14, 2018, a troubled 19-year-old student who had been expelled from the Marjory Stoneman Douglas High School in Parkland, Florida, committed a mass shooting there, using a legally-purchased semi-automatic rifle and multiple magazines. In six minutes, this former student (who shall here remain nameless) killed 14 students and three staff members and injured 17 others. In the aftermath of the shooting, student survivors organized the activist group Never Again MSD. They criticized politicians for offering weepy condolences—while dodging the issue of gun control and continuing to accept campaign funding from the National Rifle Association. The student activists demanded legislation that would produce stricter gun control measures and initiated a national protest against rampant gun violence that has become tragically familiar in the USA.

STS to FD:

1 May 2018

Dear Professor Dyson,

On behalf of the Spring 2018 semester's STS class, I bring you greetings and good wishes...

If we would not be imposing too much, the Spring 2018 STS students would like to ask you a few questions.... you may be interested to know what the students are thinking.

1. There has been much discussion and little action about gun control in the USA. We admire the Parkland, FL students who have taken a stand and articulated it effectively. Being familiar with the cultures and traditions and constitutions of various countries, where do you see the issue of gun control going in this country?

FD to STS:

4 May 2018

Dear Dwight and students,

Thank you for your message with the new list of questions. I am glad to see that nine of you are majoring in elementary and early education. I was lucky to have had a great elementary school teacher whom I remember vividly after almost ninety years. We called her Miss Scott and she taught us more than anyone else. For me the hours that I spent sitting in class in high-school were mostly a waste of time, and the time I spent with Miss Scott was much more important. I hope those of you who become elementary teachers will enjoy teaching as much as she did.

Now for the questions.

1. Gun control in the USA. I have to confess that I never understood and still do not understand the American craziness about guns. I grew up in England where very few people have guns and guns are not a big problem. Policemen do not normally carry guns, and criminals also do not normally carry guns. The habit of not carrying guns is maintained by the legal system. If anyone commits a crime and is caught carrying a gun, the sentence is much more severe than it is for the same crime if the offender is not carrying a gun. The system is friendly to criminals without guns. That seems to me a good system, better for the criminals and also better for the police and better for the public.

I find the American attitude to guns totally absurd and extremely harmful. I do not see any cure except a real change of attitude in a majority of the population. I admire the Parkland students for their courage and their eloquence and their common sense, and I hope they will succeed in their effort to change the gun control laws. But a real improvement in the American gun-laws will require much more radical changes in the American gun-culture.

STS to FD:

2. Several semesters ago, we asked you if you have a question you would like to ask us. That exchange resulted in meaningful discussions here. Do you have a question for us at this time?

FD to STS:

2. I choose a question that does not overlap with the questions 3, 4, 5 that are already on your list. I have a friend who is a Mormon and has raised a splendid family of seven kids. Each of the kids spends two years as a missionary immersed in a foreign culture and becoming fluent in a foreign language. Now the kids are grown up and the parents are doing two more years as missionaries in the Philippines. They all agree that the years of missionary work are wonderful for their own education, and not so much for making converts. In addition to educating themselves, they also provide practical help to the local people, teaching and nursing and building schools and public-health clinics. Now comes the question. Why do we leave this activity to the Mormons? Any group of people with good organization and fund-raising ability could do it. It does not need to be based on religion. It would be a great benefit to the United States if a larger fraction of our population were

> *spending a couple of years abroad to understand the world outside. How many of you are actually planning to spend a couple of years abroad, either in business activities or as missionaries?*

Our university sponsors modest summer mission trips of a few week's duration. These trips typically involve about 150 students traveling to various domestic and international sites where they are of service in various ways. Called the Students in Missions Service, these SIMS expeditions proceed with a spirit of goodwill similar to that of the young Mormon missionaries, but without the depth of immersion in another culture or achieving fluency in the local language.[3]

At the time of this STS-Dyson letter exchange, the US-Mexico border was deemed by many US voters to be too porous to refugees. Most refugees were desperate families seeking to escape dire poverty and/or threats of violence from gangs and drug cartels.

STS to FD:

> 3. Where do you see the refugee and immigration issues going? What principles should guide policy decisions?

FD to STS:

> *3. The refugee and immigration issues are difficult and will remain difficult. We have to find some compromise between harsh and friendly. At the moment the USA is one of the harshest countries, carrying far less than our share of the burden of caring for refugees. I hope we can do better in future. Two immediate improvements could be made.*
>
> *(1) Make the laws simple and clear so that refugees know whether they have a chance of being accepted.*
>
> *(2) Make decisions quickly so that detainees can get on with their lives. In the long run, we need to fix a rate of immigration that the country can handle peacefully, and then stick to it. My personal choice for immigration-rate would be around one-half per cent per year.*

STS to FD:

4. Presidents come and go, but the present [Trump] administration seems to have little human empathy, and shows lack of respect for objective reality. What could be the long-term consequences for this period in public life? Is this a trend or an anomaly?

FD to STS:

4. I have a profound respect for the American Constitution, which was designed to ensure that a government can be run by crooks without disaster. The Constitution is now being put to a severe test, but I think it will do the job it was designed for. It limits the amount of harm that any single bunch of crooks can do. In the two centuries that the Constitution has been working, the role of the judiciary has been strengthened, so that the courts have more power today than they had originally. That increases my confidence that the harm that one President can do will be limited. I find it encouraging that the prosecutors, now taking the lead in questioning the crooks, are not intimidated.

Of course I also rely on the good sense of the American voters to get rid of crooks from time to time. I have vivid memories of the 1948 election, just after I arrived in America, when we all expected Truman to lose, and to our amazement he won. I would not be surprised if something similar happens in 2020. I do not try to predict who the next Truman will be, but I trust the voters not to make the same mistake twice.

STS to FD:

5. God forbid it ever happens, but what could be the cause of World War 3, and between whom might it be fought? This question may have been motivated by recent heated rhetoric from the US President regarding North Korea, and the increasing demands made on natural resources in the light of disparity between haves and have-nots.

Thank you Professor Dyson for being part of our STS class.... Please give our greetings and good wishes to Imme and to your children and grandchildren.

I am gathering some sample students essays to send you. No reply will be necessary; they are merely for your information, a glimpse of student thoughts. You have given us so much, and we would like to give something back.

Warm regards, with deep appreciation, The Spring 2018 STS class

FD to STS:

5. I have nothing new to say about World War 3. The way to avoid World War 3 is to recognize that we are sharing the world with China and other countries. We should always talk with them, but we should not be telling them how to behave.

I remember from my school days in England learning about the War of Jenkins's Ear. The war was fought between England and Spain after the English sailor Jenkins lost an ear in a quarrel with a Spanish sea-captain. England of course won. Wars have been fought for all sorts of silly reasons. World War One is the prime example of a big war fought for silly reasons. World War Two is the prime example of a big war fought for natural resources. Both Germany and Japan went to war to grab land and oil to which they thought they were entitled. In the future it is likely that we will continue to fight wars for a variety of reasons. Attempts to abolish war by international agreements have not been successful.

In my opinion, we must plan to live in a world with wars occurring from time to time. There are two things that we have to do to improve our chances of survival.

(1) Get rid of the most destructive weapons, particularly nuclear and biological weapons. In the past we have got rid of big quantities of weapons unilaterally rather than by negotiation. Unilateral discarding of weapons is likely to be the best way also in the future. President Nixon got rid of a hundred percent of our biological weapons, and President George Bush senior got rid of fifty percent of our nuclear weapons, both by unilateral action. Unilateral action gives a clear signal to our enemies that we consider the weapons more dangerous than useful.

(2) Talk to our enemies as much as possible, especially after fighting starts. Scientists can be helpful here, because we are a

working international enterprise, and it is much easier for us to talk to our enemies than it is for politicians. A good example in recent times is Siegfried Hecker, retired director of the Los Alamos Weapons Laboratory. Hecker is our leading weapons expert. He went six times to North Korea to talk with the North Korean weapons experts. Both governments allowed these conversations because they both benefited from understanding their enemies. And in case we start fighting, personal contact between the experts will make it much easier to stop. That is all I have to say to answer your questions.

Good wishes to all of you, yours ever, Freeman Dyson

STS to FD:

5 December 2018

Dear Professor Dyson,

The Fall 2018 Science, Technology, and Society class at SNU hopes that 2018 has been a good year for you, and that this December finds you well and surrounded by family and friends.... Please give our regards to Imme and pass along our season's greetings to your family.

In recent years several students have been spending each semester at SNU's field station in Costa Rica. The Dyson Travel Scholarship that greatly helps defray their logistical expenses has been an enormous blessing. We have a weekly colloquium which features many presentations by students on their undergraduate research projects. Several presentations this semester have described work done at the Costa Rica field station, the Quetzal Education & Research Center in the San Gerardo de Dota valley in the Talamanca Mountains. Your influence has been significant and it continues.

If you have a moment to answer two or three of the following questions, we would be grateful....

1. We congratulate you on your approaching 95th birthday. If any of us are able to make it to 95, we hope we can be as vigorous in our nineties as you are. What is the view from 95? (We asked similar questions in the past, and

are wondering if you have anything to add to your former responses.)

FD to STS:

8 December 2018

Dear Dwight,

Thanks to you and the students for another bunch of questions. I am delighted to hear that they are spending time in Costa Rica and learning a bit about the world outside the USA. Here are some remarks responding to your questions. As usual, I do not answer the questions but give you some thoughts and opinions suggested by the questions.

1. The main effect of reaching 95 is a feeling of detachment. I suppose it is what people mean when they speak of the serenity of old age. I am grateful for it. I can relax. I have done what I could, and now it is time for young people like you to take over the world and try to do better. I see my tribe of grandchildren growing up, eager and hopeful, and I feel some pride. They are my contribution to history. I will not be around to see what they will do, but I am proud to have given them a good start.

STS to FD:

2. In a former letter you suggested that fossil fuels will remain the dominant energy source for the next century, but on a longer timescale society will come to rely on renewable sources such as solar energy. What cultural disruptions do you anticipate will occur as societies wean themselves off of fossil fuels? In other words, how should societies prepare for this inevitable transition?

FD to STS:

2. I do not answer your question but discuss another piece of the climate change problem. I was in California for several months in the last year breathing smoke from the big fires. One of my daughters lives in Redding and had to leave her home for three days. She was lucky. When she came back, her home was still there. Many homes nearby were destroyed and still smoking. We do not know whether these big fires have anything to do with carbon dioxide, but I see a possible link. Carbon dioxide in the atmosphere is a substitute for water. Every plant has little pores called stomata in its leaves. To grow, every plant has to open its stomata so that carbon dioxide molecules can come in from outside, and at the same time water molecules move out through the pores and are lost. If the air is enriched in carbon dioxide, less water is needed in the plant for each ton of plant growth. In a dry climate like California, increase of carbon dioxide causes growth to be faster and plants to be dryer. Faster growth and dryer plants mean bigger fires. The same rules also apply to agricultural food-crops. In a dry climate, more carbon dioxide gives us higher food-yields using less water.... So the rising carbon dioxide produces more food and also produces bigger fires. Humans must balance the costs of bigger fires against the benefits of bigger yields of food. There is no such thing as a free lunch. But we can reduce the damage from fires substantially by letting small fires burn up the vegetation frequently. We should let many small fires burn under control, instead of having a few big fires out of control.

STS to FD:

3. What are your thoughts on the recent Intergovernmental Panel on Climate Change Fifth Assessment Report?[4]

FD to STS:

3. The IPCC is by statute required to report on the effects of human activities on climate. It is not supposed to report on two other equally important subjects, the non-climate effects of human activities on plant growth including food-crops, and the effects of non-human actors such as the sun on climate. So the report is grossly one-sided, discussing only one side of a three-sided problem. In my opinion, the IPCC process is driven by politics and not by science. By limiting the subject of the report, the IPCC is pre-determining the recommendations. The IPCC report is misleading the public, misrepresenting political judgments as scientific facts.

STS to FD:

4. Of the Manhattan Project leaders you knew personally, how did their thoughts and attitudes about nuclear weapons change (or not change) between the start of the Project and its aftermath?

FD to STS:

4. I knew the Manhattan Project leaders only several years after the project ended. I have no first-hand information about their thoughts when the project started, except for Dick Feynman, who talked at length about his experiences during the war. Dick said he went into the project reluctantly, but accepted it as a patriotic duty to help his country fight a serious war. Then, after he arrived at Los Alamos, he found himself engaged completely in the technical challenges of bomb-design. He found himself enjoying the work and the company of a crowd of gifted people working together. He found he was himself unexpectedly effective as an organizer of the project. He found he was dangerously in love with bombs and with bomb-designing. As a result, he decided after the war was over to give up his security clearances, and he would never again be involved with military projects of any kind.[5]

STS to FD:

5. What are your thoughts on "multiverses?" Are they more than speculation?

FD to STS:

5. I am not interested in multiverses. I find it much more interesting to explore the one universe that we can see and touch. That does not mean that I consider multiverses to be nonsense. Multiverses may well exist. I only say that I find it uninteresting to study them unless we find some clear evidence of their existence. This is my personal preference, not a scientific conclusion. I would not discourage anyone who enjoys it from building theories of multiverses. Science is unpredictable, and multiverses might turn out to be real.

STS to FD:

6. What science problems that you once thought would have been solved by now still remain to be solved?

We thank you again, and wish you a Happy Birthday and a joyous Christmas season.

-The Fall 2018 STS class

FD to STS:

6. Problems that I expected to be solved: Understanding why ice ages happen. Understanding the earth's magnetic field. Understanding the sun's eleven year cycle of sunspot activity. Problems that I did not expect to be solved: Understanding the origin of life. Understanding the working of human memory. Understanding the working of human diseases such as cancer and Alzheimer's. But science is unpredictable and big surprises often happen. It would not be surprising if most of my expectations turn out to be wrong.

Happy Holidays to all of you, and thank you for staying in touch.

Yours ever, Freeman Dyson

STS to FD:

10 December 2018

Dear Professor Dyson,

Thank you once again for responding to us with such gracious attention. In a society whose popular culture seems to worship youth, we are so grateful to have a "tribal elder" share his wisdom that comes from a life well lived. May you reach 100 still vigorous.... On behalf of the class (who sent cards), I wish you a Happy Birthday and a joyous Christmas season....

Thank you again

Dwight, for STS students

22 Listening to Almustafa

And seeing the multitude, he went up into a mountain: and when he was set, his disciples came unto him:
And he opened his mouth and taught them, saying, ... —*Matthew 5:1*

An unexpected consequence of the Singapore conference appeared in our emails early in 2019:

> 14 January 2019
> Dear Prof. Dyson and Prof. Neuenschwander,
> Hope you had a good start to the new year!
> We are delighted to inform you that we are publishing a new journal *The Physics Educator*. It will be co-published with the Institute of Physics Singapore. It will be wonderful if you could contribute an article on the STS class to the inaugural issue that is scheduled for March 2019, or if the time is too tight, maybe the next in June 2019.
> Looking forward to hearing from you!
> Best regards, Lakshmi Narayanan (Editor)

FD to Lakshmi (cc: DN):

> *15 January 2019*
> *Dear Lakshmi,*
> *Thank you for your message and the invitation to collaborate with Dwight Neuenschwander in writing an article about his class.... If the article is written, Dwight must write it. I will wait to see how he decides. If he decides to write the article, I will be glad to add comments and details that I happen to remember. If he decides not to write it, I will also say no.*
> *All good wishes to your new journal and to you personally.*
> *Yours ever, Freeman Dyson*

DN to Lakshmi (cc: FD):

15 January 2019

Dear Lakshmi,

Congratulations on the new journal, *The Physics Educator*. It would be an honor to work with Professor Dyson and World Scientific and you on behalf of the STS students, in contributing an article to this new journal.

I appreciate Professor Dyson's response to you which he made this afternoon. Any comments and details he might add to whatever I can put together would give the article an immediacy and authenticity that will go beyond anything I can say....

Thanks to you, World Scientific, and Professor Dyson for this opportunity to share in another way the life lessons we have learned through Professor Dyson's patient and gracious correspondence with us.

Warm regards, and Happy New Year to you too,

Dwight E. Neuenschwander

FD to LN and DN:

15 January 2019

Dear Lakshmi and Dwight,

Dwight's enthusiastic acceptance takes me by surprise. I am delighted that the project to write the article will be going ahead on a rapid timescale. Unfortunately, this happens to be a particularly difficult time for me to meet the March deadline. I am now at San Diego working full-time for JASON, the only time in the year when I am gainfully employed. I will be in San Diego until the middle of March... After I return to Princeton in March... I will have the busiest time of the year catching up with Institute activities.... I could be most helpful if Dwight would send me direct suggestions of questions to answer and particular episodes to describe...

Yours ever, Freeman

At this point I felt like a musician who has only performed local gigs but is suddenly asked to share the stage with a virtuoso of international fame. How to approach this? Long walks facilitate thinking. On such a walk my mind turned to Kahlil Gibran's masterpiece *The Prophet*.[1] The prophet Almustafa, "the chosen and the beloved," is preparing to depart the city of Orphalese where he has lived with its people for twelve years. Sad to see him go, the citizens gather at the dock to see him off. As they await the ship's arrival, to glean his wisdom through the last possible moment the Orphalese citizens ask Almustafa questions about life: "Tell us of Giving; tell us of Joy and Sorrow; tell us of Freedom; tell us of Reason and Passion; tell us of Friendship" and so on. For almost thirty years Professor Dyson had been our Almustafa. I suggested to him that our article follow a motif similar to Gibran's. With their questions, this plan would make the Spring 2019 STS class part of the article too. The class sent two very broad questions to our Almustafa. Professor Dyson acknowledged receiving them:

FD to DN:

16 April 2019

Dear Dwight,

.....I share your admiration for The Prophet, and I will be glad to follow your suggested two questions in the style of Almustafa..... I hope to have a text with answers to the two questions within a week or two. I am still feeble but visibly improving as the weeks go by. Yours, Freeman

Within a few days he returned his answers:

FD to DN:

20 April 2019

Dear Dwight, I send you this draft to use in any way you find appropriate. Please feel free to edit, shorten or rearrange it to fit in with the rest of the article....

Yours ever, Freeman

We now had everything for the article. It remained to thank him for his answers and assemble the final draft.

DN to FD:

23 April 2019

Dear Professor Dyson,

Thank you, your replies are beautifully thought out and written. I am copying the students… I'll work your responses into the draft, and will send you the result soon for your final approval before sending the final version to *The Physics Educator*.

Lakshmi requested some lines about "interactive learning." It seems to me that corresponding with a leading participant and witness to so many of the important and exciting developments in science over the last few decades is a splendid form of interactive learning.

It is an honor and a privilege to participate in this project with you….

Warm regards to you and Imme and your family. Your grandchildren are lucky to have you as their grandfather.

Dwight

A week later the draft was completed.

FD to DN and STS:

30 April 2019

Dear Dwight,

Thank you for sending the draft of the Physics Educator article. I was surprised to see that you left the text of my answers to the students almost completely unchanged. I am grateful to you for giving me so much of the space and taking so little for yourself. I am happy with your draft of the article and do not have any further changes to suggest.

All is well with me and the family. We have been enjoying a visit from daughter Mia and granddaughter Bryn. Mia is the Presbyterian minister who has a problem because the income from her ministry is much less than she needs to pay for her kids' college education. She has solved the problem by starting a successful business as a dog-breeder, raising pedigree pups which she sells to wealthy owners for exorbitant prices. Bryn has just finished law-school and passed the bar examination so she can practise law in Massachusetts, following the footsteps of my mother who practised law in England a hundred years earlier. So now we are ready to sue if anyone steps on our turf.

Thanks again to you and the students. Yours ever, Freeman

Here, with the publisher's permission, is the Dyson-STS collaborative paper as it was published.[2]

Youth Engaging Almustafa:
Cross-Generational Interactions in
"Science, Technology, and Society" Education

Freeman J. Dyson, Institute for Advanced Study, Princeton, NJ, USA;
Dwight E. Neuenschwander & many students, Southern Nazarene University, Bethany, OK, USA
The Physics Educator, Vol. 2, Issue 2, June 2019

For more than sixty semesters, over two thousand university juniors and seniors have taken our general-education course "Science, Technology, and Society" (STS). This vast subject stands at the intersection where human ingenuity and appetites meet nature's realities. STS does not recite answers to complex questions. Rather, we attempt to understand the questions.

STS inquiries range from the meaning of evidence-based reasoning to asking what we own, what owns us, and what we borrow from our grandchildren. Topics range from nuclear weapons to astronomical habitat; from genetic manipulation to environmental stewardship; from the freedoms that automobiles and social media offer to the controlling tentacles they produce. The underlying motifs are *appreciation* and *awareness*. We should appreciate what science and technology offer, while being aware of their hidden costs. Ultimately, STS is about human values expressed in our relationships with nature, our machines, other cultures, other species, and personal identity.

Inquiry-based learning means interactive engagement. For example, to experience evidence-based reasoning on which science depends, STS students estimate the number of hay bales or ping-pong balls that could be packed into a campus building. Such activities require making conceptual representations of systems, recognizing assumptions, and estimating uncertainties. Elsewhere, automotive history discussion begins with students dissecting lawn mower motors, to see the internal operation of machines on which they depend.

A priceless mode of inquiry-based learning seeks conversations with elders. Robust societies need meaningful interactions between elders and the young. From elders, the young glimpse memories they did not live, and a depth of experience they have yet to acquire. From the young, elders are rejuvenated with questions, ideals, and an enthusiasm for life that looks to the future.

From the beginning of STS in 1986, Freeman Dyson's memoir *Disturbing the Universe* has been our primary textbook. In its opening lines he observes "The best way to approach the ethical problems associated with science is to study real dilemmas faced by real scientists."(1) Still writing and consulting in his mid-nineties, his life experiences and professional accomplishments, graced by his authenticity and humility, cast a long shadow. Since the Spring 1993 semester our STS classes have corresponded with Professor Dyson.(2) He has been a wise grandfatherly mentor to two generations of university undergraduates. We are reminded of Almustafa in Kahlil Gibran's poetic masterpiece, *The Prophet,* which opens with Almustafa awaiting his ship. The citizens of Orphalese gather around him, saying "You have walked among us in spirit, and your shadow has been a light upon our faces... Yet we ask...that you speak to us and give us of your truth."(3) With respectful entreaties they say "Tell us of Freedom" and "speak to us of Reason and Passion,... of Good and Evil,... of Self-Knowledge,... of Children...." In response, Almustafa illuminates the parameters of deep questions for which answers are elusive. Sometimes it is enough to appreciate the questions.

Now the Spring 2019 STS class has walked with Professor Dyson through *Disturbing the Universe*. Like the citizens of Orphalese entreating Almustafa, again we humbly seek Professor Dyson's insights; and like those citizen's questions ours are broad ones, about humanity's struggle to achieve lofty goals. "Dear Professor Dyson, 'you have walked among us in spirit, and your shadow has been a light upon our faces.' Please speak to us of Science and Technology, Social Justice and Civic Awareness." He answered,(4)

"Social justice means a fair share of the necessities of life, and of the opportunities for creative leadership, to every child. In the words of Thomas Jefferson, social justice means that we are all created equal. We are free to use our opportunities well or badly, to earn wealth by making big contributions to society, to become poor by failing to contribute. A socially just society does not abolish inequality, but it keeps inequality within limits that we accept as fair, treating rich and poor with equal respect.

"Social justice belongs to the laws and institutions of a society as a whole. Civic awareness is a goal for individuals. Civic awareness is the set of emotions and loyalties that make membership of a group the driving force in the life of an individual. Humans evolved as social animals, compelled to work and think in groups by the necessities of a hunting life-style. Civic awareness was long ago written into our genes. Our recent cultural evolution has transferred civic awareness from villages and tribes to nation-states and religions. Now we hope for a future transfer to our species as a whole, to a peaceful brotherhood of humans working together.

"You ask how our social goals are related to science and technology. Here is the answer. Science and technology are a set of tools devised for the understanding of nature, which automatically become available to humans everywhere as soon as they are invented anywhere. Science and technology are inherently international. They are the strongest forces pulling our species

together. They pull us into a world-wide community of thinkers and builders who share ideas and skills. That is why science and technology must be an essential part of the education of every child. The purpose of scientific education is not to force every child to become a scientist. The purpose is to give every child the awareness of an international community already working together with common goals and shared benefits. Every scientifically educated citizen knows that friendships and collaborations reaching around the world are possible."

"Professor Dyson, please tell us of Science, Technology, and Environmental Responsibility." He answered,

"The goal of environmental responsibility belongs both to individuals and to societies. The aim is to regulate the society and our own activities so as to protect the marvelous diversity of the natural environment. Environmental responsibility is difficult because of evolution. Both nature and the human species are constantly evolving. Neither could survive without the creative force of evolution to overcome disasters and replace worn-out parts. Environmental responsibility does not mean trying to stop the environment from changing, or to stop nature and the human species from evolving. To stop evolution would be fatal to both. Environmental responsibility must be based on compromise, reconciling the demands of natural and human evolution where they come into conflict. Natural and human evolution are both high-risk strategies. Nature evolves by forming new species prolifically with high risk of failure. Humans evolve by forming new institutions and enterprises with high risk of failure. Neither nature nor humans would do better with a strategy of risk-avoidance. The outcome of environmental responsibility should be a dynamic interplay of creative diversity of new species with creative diversity of human life-styles. Neither nature nor humans can survive by standing still.

"One of the most important achievements of human evolution was the invention of grandparents. After we began living in permanent settlements, the post-menopausal life-time of women

lengthened. Men were also living longer after their hunting skills declined. Women and men who survived to old age became grandparents, a third generation who raised and educated grandchildren while parents were hunting and gathering. After we invented spoken language, grandparents became far more effective educators, transmitting knowledge and memory from generation to generation. Grandparents, telling stories to grandchildren sitting around the cave-fire, became the creators and guardians of a permanent culture. Culture was tribal tradition and mythology, rituals and religions, music and poetry.

"Latest of all in the evolution of culture came science and technology. Science and technology are now the main driving forces of our continuing evolution, changing us from a mob of warring tribes to an emerging international community with hopes of world-wide human brotherhood. I am proud to be a grandparent, helping to educate students together with my own grandchildren, sharing with them my vision of the promises and dangers of science and technology. I try to give them wisdom to achieve the promises and mitigate the dangers."

The Lakota elder Joseph Marshall III observed, "We live in a world that moves at cyber speed, craves instant gratification, and revels in technology. Consequently, we are so impressed with the current version of ourselves that we aren't aware that our ancestors contributed to what we are and what we do and how we think."(5) For lessons in appreciation, awareness, and how to *care,* we value the wisdom of our predecessors. Professor Dyson has been our mentor, generously sharing with us his deep wisdom "to achieve the promise and mitigate the dangers." "What was given us here we shall keep."(6)

Acknowledgments

The Spring 2019 STS class that sent the questions in this article are: Deanne Adams, Mitchell Aldridge, Jumanah Alkhanizi, Jon Beckman, Luka Belamaric, Madeline Dixon, Eric Dyrssen, Clayton Elliott, Megan Ellis, Emily Ferrell, Rebecca Janka, Hema Joseph, Nancy Jurado, Zachary

Overstreet, Jorden Richard, Easton Rodgers, Crystal Roopnarinesingh, Melody Ryan, Nadine Samilpa, Niki Spohn, Kayla Thomas, and Kaleb Yost. On behalf of two generations of STS students who have been mentored by Professor Dyson, we express to him our deep gratitude.

Other elders with whom we have consulted on occasion include Kenneth Ford, who as a graduate student worked on the first hydrogen bomb; and Lenard Hauk, the photographer aboard the B-29 that followed the *Enola Gay* on the Hiroshima mission.

(1) Freeman Dyson, *Disturbing the Universe* (Basic Books, New York, 1979), p. 6.

(2) For a collage of STS discussions and the 1993–2016 correspondence with Professor Dyson, see *Dear Professor Dyson: Twenty Years of Correspondence Between Freeman Dyson and Undergraduate Students on Science, Technology, Society and Life*, Dwight E. Neuenschwander, ed. (World Scientific, Singapore, 2016).

(3) Kahlil Gibran, *The Prophet* (Alfred Knopf, New York, 1964), pp. 8, 10.

(4) The questions were E-mailed to Professor Dyson, with the Spring 2019 semester's STS student's names attached, on April 16, 2019. His reply arrived on April 20.

(5) Joseph Marshall III, *Walking with Grandfather* (Sounds True, Boulder, CO, 2005), pp. 3–4.

(6) Ref. 3, p. 94.

The Spring 2019 class, who participated in the "Almustafa" article for The Physics Educator (DN photo).

23 Doubt, Faith, and Peaceful Coexistence

There lives more faith in honest doubt,
Believe me, than in half the creeds.
—Alfred, Lord Tennyson, *In Memoriam*

Our STS classes began corresponding with Professor Dyson during the year when he turned seventy years of age. Every time we corresponded with him we always felt fortunate that his ship had not yet come for him. Through the wisdom and person of Professor Dyson, we were given a great gift. We never took Professor Dyson's gift for granted.

DN to FD:

13 September 2019

Dear Professor Dyson,

It was an honor to work with you on the article for *The Physics Educator*. Thank you for the privilege of working with you.

We are now about three weeks into a new semester, with another class of Science, Technology, and Society. One student wrote in his weekly essay, "Something that really stuck out to me in this reading of Dyson...was about his family's social status. Professor Dyson says 'I was gifted with brains, good health, books, education, a loving family, not to mention food, clothing, and shelter...." [*DU* p. 17] This student goes on to describe how "I have taken for granted my family's wealth and social status... I've complained about my bed when some people don't have one. I've complained about my family's dinner when people don't have either a family or money to pay for dinner... So lately I have been trying to institute small daily practices in my life to help me get a different and more aware perspective of what I have. Such as only sleeping in my bed on the weekends and sleeping on the floor during the week...."

Your influence on the students here reaches well beyond issues of science and technology....

If you don't mind hearing from another STS class, we will plan to send you a few questions near the end of the semester. We do not want to trade on your time or take you for granted, so do not feel obliged to answer us. But we want you to know that you have many fans here. -Dwight

FD to DN:

13 September 2019

Dear Dwight,

Thank you for telling me about your student sleeping on the floor. That is a response I would never have imagined. You can tell him or her that I never did that myself, and I am delighted to hear about it....

Yes, I will be glad to hear from the students any time and try to answer their questions....

My greetings and good wishes to the students and to you. Yours ever, Freeman

STS to FD:

3 December 2019

Dear Professor Dyson,

On behalf of the Fall 2019 semester STS class at SNU, we hope you and your family enjoyed a wonderful Thanksgiving holiday and are anticipating a splendid Christmas season with family and friends. If we may, the class would like to ask you a few questions.... But please feel under no obligation. We do not want a moment of your time with us to come at the expense of time with your family.

1. What was the hardest lesson you learned growing up, and how did it change your outlook on life? This question was stimulated by our discussions of how we as individuals, and as a society, learn more from challenge and failure than we do from ease and success.

FD to STS:

10 December 2019

Dear Dwight and students,

Thank you for another bunch of questions. I will try to find answers that tell you something new.

1. The hardest lesson from growing up. When I was twelve years old I was the youngest kid in my high-school class and also got the highest grades. I considered it unfortunate that the grades were not publicly announced. So I compiled a list of the grades with my name at the top, and posted the list on the class-room notice-board. Soon after that, the class teacher, who happened to be a Christian pastor, called me in for a serious talk. He pointed out that my presence in the class was already a cause of resentment, and if I continued to call attention to myself I would soon turn all my friends into enemies. He said that, if I did not learn to treat my less gifted colleagues with respect, he would be compelled to have me removed from his class. His name was the Reverend Dick David, and his quiet words have remained as a voice of wisdom for the rest of my life.

STS to FD:

2. We have seen your responses to previous classes about long-term energy sources, such as how fossil fuels will probably dominate for another century, with solar and wind power becoming abundant and cheap on that timescale.... How do you envision society meeting its energy needs, say, two centuries from now? We can imagine solar panels and small wind turbines on the roof of every household and commercial building, instead of vast landscapes covered by turbine farms and solar panels. What do you envision? As part of this question, what might be the long-term role of nuclear power? We read how in the 1950s the petroleum geologist M. King Hubbert predicted that nuclear power would dominate after fossil fuels are gone.[1]

FD to STS:

2. My guess for the future of energy supply in the long run is solar. Solar energy is abundant, well distributed, clean, and in the long run cheap. It has been Nature's choice for the driving force of life for billions of years. The question to my mind is only how long it will take to solve the storage problem, to invent a way of storing energy that is cheaper and safer than electric batteries. Nature's way of storing energy uses chemical fuels such as cellulose and animal fat. I imagine that we shall learn to imitate nature by biological engineering of trees and microbes on a massive scale. I would be surprised if this takes us longer than two hundred years or shorter than a hundred years, to grow into a cheap and efficient world-wide energy storage system. But the success or failure of new technologies is never predictable. Nuclear power is a glaring example of an unexpected failure.

STS to FD:

3. We deeply appreciate your "two windows" description of the relation between science and religion. That mental picture is very illuminating and helpful. Amid that topic's discussion we examined Bertrand Russell's definition of faith ("a belief one holds in spite of evidence to the contrary" from *Why I Am Not a Christian*)[2] and the definition offered by Walter Kaufmann ("a belief one holds when there is not sufficient evidence to compel all reasonable observers to agree" from *The Faith of a Heretic*—Kaufmann also observed that "certainty should not be purchased at the price of honesty").[3] In your view, Professor Dyson, what is the optimal relationship between doubt and faith?

FD to STS:

3. The optimum relationship between doubt and faith is peaceful coexistence. Both are essential to the evolution of a creative human society. Faith to pursue impossible goals, doubt to recover from disastrous mistakes. We have to learn to tolerate a wide variety of faiths and doubts. Hope for a better future lies in the diversity of tyrants and rebels that our evolution brings to life.

STS to FD:

4. In Ken Burns' documentary *The Civil War*,[4] Burns tells how a young man asked Frederick Douglass for advice on what he should do with his life. Burns reports that Douglass replied, "Agitate, agitate, agitate." If a young person were to ask you a question similar to the one put to Frederick Douglass, how would you answer?

FD to STS:

4. Humans are born with a wonderful diversity of talents and tastes. Some are born to be agitators, others to be explorers, others to be teachers or artists. I find the advice given by Frederick Douglass far too narrow. Each of us should be urged to contribute his or her particular gift to the goals of social progress or justice. Douglass probably knew the people he was urging to agitate, so his advice was very likely appropriate for that audience.

STS to FD:

5. In a previous letter you described how the problems of social injustice are more urgent and important than the energy problem. If a young person is passionate about usefully addressing problems of social injustice, what causes or activities would you suggest they prioritize?

FD to STS:

5. This is the same question as number 4 and has the same answer. The problems of social injustice are deeply embedded in societies all over the world, perpetuated by wealthy families and powerful politicians and racial prejudices and residential zoning laws and unequal access to jobs. The remedies must be found at all levels, within families and institutions and religions and taxation systems. The quickest remedy would be a massive shift of taxation from poor to rich. To achieve such a shift, advances in science and technology may be helpful or harmful. The most effective way to move forward would be to combine the tools provided by science and ethics and law and religion.

STS to FD:

6. All science applications expressed in technology are subject to what historian Theodore White called the "law of unintended consequences."[5] For example, when college students invented Facebook to keep in touch with classmates, they probably did not envision social media becoming weaponized. While no one can fully anticipate the unintended and sometimes harmful uses that may be put to a scientific discovery or new technology—as you pointed out in Chapter 1 of *Disturbing the Universe*—what *can* scientific communities do to mitigate deleterious unintended consequences of their work?

Thank you again Professor Dyson for your gracious interactions with our classes across the years.... We wish Season's Greetings to you and your family, and we also wish you a Happy Birthday which is coming up soon.

Warm regards, The Fall 2019 STS class

FD to STS:

6. The founders of the United States showed unusual wisdom in their design of the United States Constitution. They deliberately designed the division of power so that the government could be run by an unscrupulous gang of crooks without disaster. The law of unintended consequences makes it inevitable that power will sometimes fall into the hands of unscrupulous crooks. The Constitution ensures that those who abuse power shamefully can be removed from office. The durability of the republic demonstrates that the Constitution works as the founders intended. Now in the last century the rapid advance of technology has resulted in unintended consequences of a similar kind. New technologies give birth to rapidly growing industries with massive databases that allow gangs of crooks to dominate economic and political affairs. Legal governments and judicial systems are powerless to control these crooks who possess superior information systems. The remedy for these unintended consequences of technology is to borrow the design of the United States Constitution. Every new technical enterprise that grows beyond a certain size must be exposed to public scrutiny. Massive databases must be shared with competing institutions. A legal process of impeachment must be available to remove the human bosses of large enterprises. Following the model of the United States founders, we may allow a free society to profit from technical progress up to a certain size, but to meet a firm limit when social enterprises become too big to control.

These answers to your questions raise many new questions which I will be glad to discuss if any of you are interested. Happy Holidays to all of you.

Yours ever, Freeman Dyson

STS to FD:

20 December 2019

Dear Professor Dyson,

Your responses of Dec. 10 to our latest batch of questions of Dec. 3 stimulated some interesting discussion. Here is a collage of some thoughts that emerged.

Everyone could relate to question No. 1, and speculated that your talk with Reverend Dick David at a young age may have been a factor in how gracious you have always been to those of us who are not Richard Feynmans or Julian Schwingers, but elementary education or business majors or teachers at a small college in a fly-over state. The students here are in awe of your intelligence and accomplishments, but what they take away is how you make time to talk authentically to students, by letter or email or around a table at a Sigma Pi Sigma Congress—advice and example from a wise grandfather trying to help us make positive differences in the world.... Your approachability and your authentic engagement with us... gives our efforts a sense of dignity and encourages us to do our best....

Thank you for your response to our question about the proper relationship between faith and doubt: "peaceful coexistence." Those of us who were raised in very conservative religious homes were often presented with a false dichotomy... By the time these students take STS in their junior or senior year, most of them have come to realize that raising questions and recognizing honest doubt is the only path to deeper understanding—or a faith that can be authentically one's own.... Your statement about faith and doubt leading a "peaceful coexistence" was a fitting benediction to the semester.

Your response to the question based on the Frederick Douglass incident was appreciated. Agitators have their place, and the world needs some of them, but not everyone can or should be an agitator.... One point made in every STS course is "You and I are not called to change the world. But we *are* called to do what we can, where we are, with whatever we have...."

Your extension of the gang of crooks theme to include the technology-empowered gang of crooks that now dominate commerce and information was an appropriate point to add to the class, where we repeatedly ask the question "Do we own our technology, or does it own us?" ... It seems that we do have a State Religion—we worship the God of Convenience. Convenience has its place, but it is not a good teacher, nor is it a builder of human agency.

I am sending you the latest financial report on the Dyson Travel Scholarship that enables students to study and carry out research projects at our Quetzal Education Research Center in Costa Rica... We trust that students will be enabled to visit the QERC for many generations to come, thanks to your generosity with us at the time of your Templeton Award. For that I thank you again on behalf of all the students past, present, and future who have benefited or will benefit from it.

By the time you read this your birthday will have come and gone for 2019, and it will be nearly Christmas or past it.... May you and your family have a wonderful holiday season. The students have gone home now for the Christmas holiday, and the campus is very quiet. I hope you had a wonderful birthday last weekend.

I wish you continued good health and a robust continuation of your thinking and writing. May you still be going strong at 100!

Warm regards and best wishes to you and to all your family,

-Dwight, on behalf of the students

24 Requiem

"It is when you give of yourself that you truly give... What was given us here we shall keep... You have walked among us a spirit, and your shadow has been a light upon our faces. Much have we loved you.... Yet this we ask ere you leave us, that you speak to us and give us of your truth. And we will give it unto our children and they unto their children, and it shall not perish."

—Kahlil Gibran, *The Prophet*

Alas, Professor Dyson would not live to see 100. On 26 February 2020 he suffered a fall at the Institute for Advanced Study. He was taken to hospital and passed peacefully away on February 28. Our Almustafa took passage on his ship.

Passing from this life is the price of admission into it. While sad, the passing of a long life well lived carries the counterweight of gratitude and blessing. We are blessed and grateful to have Professor Dyson be our mentor for almost 30 of his 96 years. We are grateful to be counted among his many friends. Much we learned from him as he taught us how to know ourselves.

Professor Dyson described babies as a little piece of God. He loved babies. *Disturbing the Universe* closes with a three-month old baby having the last word—in the form of a smile. If there were no death, there would be no need for babies. But who would want to live in a world without babies! In the awesome and mysterious journey of life, as Professor Dyson has said, "holding babies is an act of worship."

The students in the Spring 2020 semester's STS class were of course deeply disappointed that they could not correspond with Professor Dyson personally. Thankfully, we have his legacy, his letters, his books, his life story, and memories of our engagements with him. "What was given us here we shall keep."

As the first anniversary of Professor Dyson's passing approached, the class of the Fall 2020 semester sent a sympathy card to Imme. From her we received a moving reply. Imme thoughtfully enclosed the Dyson family's lovely 2020 Christmas card, a message of warmth, and an eloquent tribute to Freeman.

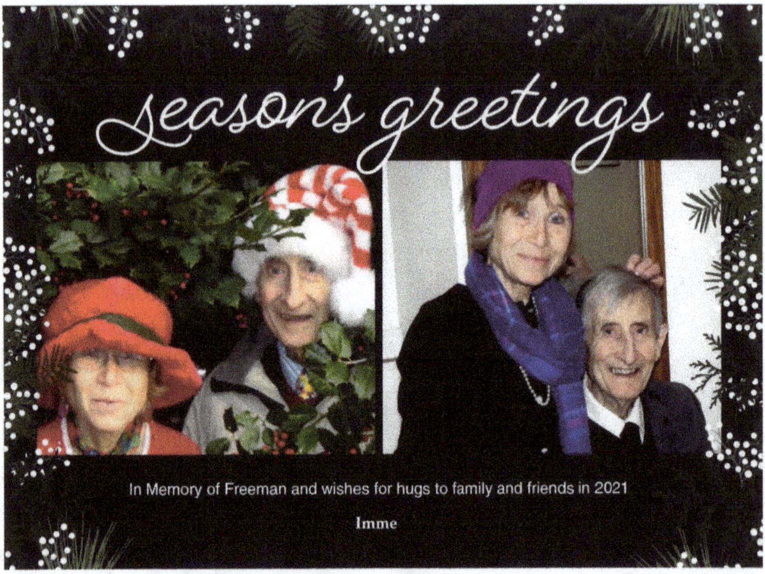

Christmas card courtesy of Imme Dyson. Thank you.

Dear Dwight, *December 23, 2020*

Needless to say, the tears had to subside before I could respond to your warm, loving memories of Freeman's friendship and correspondence with you and your students. Thank you!

Freeman loved young people, he saw the future in them, he loved their eagerness to question and watch their minds grow. As the time grew near, he would lift his eyebrows, look across the breakfast table, and say "Any day now, I am expecting the student's questions from Nazarene." It was always a good day for him.

I don't remember the incident with the grandchildren interrupting a formality, but it would be so like Freeman to give them priority! It is very sad that he was denied the greatest joy of greeting his first great grandchild, Harrison Freeman. He arrived exactly 8 months after Freeman died. The rest of the Dyson clan is rallying around him like an elephant herd over a new arrival!

With much love and hopes for an exciting future to the students,

Yours ever (in Freeman's spirit)

Imme

Professor Freeman J. Dyson. Farewell, our friend. Our hearts are filled with gratitude. Photo courtesy of the Dyson family.

Doth not wisdom cry? and understanding put forth her voice?...
Hear: for I will speak of excellent things; and in the opening of my lips shall be right things.
For my mouth shall speak truth…
All the words of my mouth are righteousness…
For wisdom is better than rubies;
And all things that may be desired are not to be compared with it.
—*Proverbs*, Chapter 8

Appendix 1

Two Views of the Future

Homily given by Professor Dyson at Gustavus Adolphus College Saint Peter, Minnesota, April 12, 1999

I don't know why you gave me the text of Jonah and the whale to preach from. I am a visitor here, enjoying the hospitality of the College, but I don't feel like Jonah. I am certainly not responsible for last year's tornado, and you people are not like the people on the ship going to Tarshish. Instead of finding a Jonah to blame for the tornado, you stood together and repaired the damage and got on with your lives. I am deeply impressed by the way you came through the disaster and took care of one another. You didn't need to find a Jonah to throw into the sea.

That's all I have to say about Jonah. I chose the text for this homily from Luke Chapter 8, verses 5 to 8. But the text will come at the end instead of at the beginning.

I want to tell you about two things I did during Spring break. I saw two very different visions of the future. The first was a meeting of about six hundred leaders of the computer and software industries. These people are mostly young, owners and CEOs of companies, some wildly successful, others hoping to be successful. Many of them bring their kids along to the meeting. There were about seventy-five kids there, including four of our grandchildren. The meeting is a get-together of people who think they are the wave of the future. They are building a new world, international, multilingual, friendly to women and children. About twenty-five percent of the owners and CEOs of companies at the meeting were women. They believe they are making the world a better place, creating jobs, creating tools for people all over the world to get ahead, to educate themselves and start new enterprises. If some of the leaders get indecently rich creating these good things, so much the better. Whether you like it or not, the world these people are building is growing fast. It is here to stay.

What do you need to live well in the new world? You don't need to be a scientist. You don't need to have a Ph.D. or even an MBA. You need to understand people. You need to write clearly, speak clearly and think clearly. You need to be able to write a business plan so that bankers will lend you money. You don't need to understand the inner workings of computers or the inner workings of people. You need to understand computers the way you understand people. You need to understand what computers can do, what networks can do, what computers and people can do together. You need to be wired. And you need to know a modest amount of mathematics. You need to know the difference between a million and a billion.

That is one vision of the future, the future seen by young, rich, upwardly mobile capitalists. Here is another vision, the future seen by people who are not rich and upwardly mobile. It comes from a book called *Parable of the Sower* by Octavia Butler.[1] In May I shall be on a television program in Chicago with Butler, debating the future of science and technology. The studio gave me three of Butler's books to read. She is a famous writer of science fiction. *Parable of the Sower* is not really science fiction. It is first-rate fiction, but there is more religion in it than science. I would call it theo-fiction rather than science fiction. Butler grew up poor and black in California, and she writes about the world she grew up in, as it might be in the future if things go badly. In the book, things go very badly. The climate has changed so that Southern California gets no rain at all. The minority of rich people live in armed fortresses. The majority outside are homeless and hopeless, scavenging in the ruins of civilization. Butler's hero is a young black woman, the daughter of a Baptist preacher, who escapes when her home is destroyed and her family is murdered. She leads a handful of survivors on a long trek to the north. As she travels, she works out a personal religion which she calls Earthseed. She says, "God is power, infinite, irresistible, inexorable, indifferent. And yet, God is pliable, trickster, teacher, chaos, clay. God exists to be shaped. God is Change". This is not the Baptist religion of her father. This is something new.

Butler's story is supposed to happen in the year 2027. I won't be around then, but you students will still be in your forties and fifties. You will be half-way through your careers, watching your children grow. Which of these two futures will be yours? The future of the Yuppies, dreaming that they can bring wealth and enlightenment to the world with high-bandwidth networks and a global market economy. Or the future of the street people, left behind by deteriorating public services and inaccessible technology. The choice is up to you. Your generation will have the power and the responsibility to decide which way to go. It's fine to get rich, but it's not fine to build fortresses to keep out the poor. Butler's story ends with the parable of the sower. That's better than Jonah as a text for my homily

"A sower went out to sow his seed, and as he sowed, some fell by the wayside, and it was trodden down, and the fowls of the air devoured it. And some fell upon a rock, and as soon as it was sprung up it withered away, because it lacked moisture. And some fell among thorns, and the thorns sprang up and choked it. And others fell on good ground, and sprang up, and bore fruit an hundredfold."

Appendix 2

Progress in Religion

Freeman J. Dyson, Institute for Advanced Study, Princeton, NJ
Lecture at Washington National Cathedral
Acceptance Speech, Templeton Award for Progress in Science and Religion
May 16, 2000

First, a big thank-you to Sir John Templeton and the administrators of the Templeton Foundation for giving me this undeserved and unexpected honor. Second, a big thank-you to the Institute for Advanced Study in Princeton for supporting me as a Professor of Physics while I strayed into other areas remote from physics. Third, a big thank-you to the editors and publishers of my books for giving me the chance to communicate with a wider public. Fourth, a big thank-you to my wife and family for keeping me from getting a swelled head. And fifth, a big thank-you to the Washington National Cathedral for allowing us to use this magnificent building for our ceremonies.

Sir John Templeton has told us clearly the purpose of his awards. They are prizes for Progress in Religion. But it is up to us to figure out what Progress in Religion means. Roughly speaking, there have been two main themes in the lives of previous prize-winners. The first theme is practical good works, caring for the poor and sick, helping the dying to die with dignity. Outstanding among the doers of good works were Mother Teresa and Dame Cicely Saunders. The second theme is scholarly study and teaching, helping people who are committed to one religion or another to approach God through intellectual understanding, explaining to the uncommitted the logical foundations of belief. Outstanding among the scholarly prize-winners are James McCord and Ian Barbour. I am amazed to find myself in the company of these great spirits, half of them saints and the other half theologians. I am neither a saint nor a theologian.

To me, good works are more important than theology. We all know that religion has been historically, and still is today, a cause of great evil as well as great good in human affairs. We have seen terrible wars and terrible persecutions conducted in the name of religion. We have also seen large numbers of people inspired by religion to lives of heroic virtue, bringing education and medical care to the poor, helping to abolish slavery and spread peace among nations. Religion amplifies the good and evil tendencies of individual souls. Religion will always remain a powerful force in the history of our species. To me, the meaning of progress in religion is simply this, that as we move from the past to the future the good works inspired by religion should more and more prevail over the evil.

Even in the gruesome history of the twentieth century, I see some evidence of progress in religion. The two individuals who epitomized the evils of our century, Adolf Hitler and Joseph Stalin, were both avowed atheists. Religion cannot be held responsible for their atrocities. And three individuals who epitomized the good, Mahatma Gandhi, Martin Luther King and Mother Teresa, were all in their different ways religious. One of the great but less famous heroes of World War Two was Andr'e Trocm'e, the Protestant pastor of the village of Le Chambon sur Lignon in France, whichsheltered and saved the lives of five thousand Jews under the noses of the Gestapo. Forty years later Pierre Sauvage, one of the Jews who was saved, recorded the story of the village in a magnificent documentary film with the title, "Weapons of the Spirit". The villagers proved that civil disobedience and passive resistance could be effective weapons, even against Hitler. Their religion gave them the courage and the discipline to stand firm. Progress in religion means that, as time goes on, religion more and more takes the side of the victims against the oppressors.

For Ian Barbour, who won the Templeton Prize last year, religion is an intellectual passion. For me it is simply a part of the human condition. Recently I visited the Imani church in Trenton because my daughter, who is a Presbyterian minister, happened to be preaching there. Imani is an inner-city church with a mostly black congregation and a black minister. The people came to church, not only to worship God, but also to have a good time. The service is informal and the singing is marvelous. While I was there they baptized seven babies, six black and one white. Each baby in turn was not merely shown to the congregation but handed around to be hugged by everybody. Sociological studies have shown that violent crimes occur less frequently in the neighborhood of Imani church than elsewhere in the inner city. After the two-hour service was over, the congregation moved into the adjoining assembly room and ate a substantial lunch. Sharing the food is to me more important than arguing about beliefs. Jesus, according to the gospels, thought so too.

I am content to be one person in the multitude of Christians who do not care much about the doctrine of the Trinity or the historical truth of the gospels. Both as a scientist and as a religious person, I am accustomed to living with uncertainty. Science is exciting because it is full of unsolved mysteries, and religion is exciting for the same reason. The greatest unsolved mysteries are the mysteries of our existence as conscious beings in a small corner of a vast universe. Why are we here? Does the universe have a purpose? Whence comes our knowledge of good and evil? These mysteries, and a hundred others like them, are beyond the reach of science. They lie on the other side of the border, within the jurisdiction of religion.

My personal theology is described in the Gifford lectures that I gave in Aberdeen in Scotland in 1985, published under the title *Infinite in All Directions*. Here is a brief summary of my thinking. The universe shows evidence of the operations of mind on three levels. The first level is elementary physical processes, as we see them when we study atoms in the laboratory. The second level is our direct human experience of our own consciousness. The third level is the universe as a whole. Atoms in the laboratory are weird stuff, behaving like active agents rather than inert substances. They make unpredictable choices between alternative possibilities according to the laws of quantum mechanics. It appears that mind, as manifested by the capacity to make choices, is to some extent inherent in every atom. The universe as a whole is also weird, with laws of nature that make it hospitable to the growth of mind. I do not make any clear distinction between mind and God. God is what mind becomes when it has passed beyond the scale of our comprehension. God may be either a world-soul or a collection of world-souls. So I am thinking that atoms and humans and God may have minds that differ in degree but not in kind. We stand, in a manner of speaking, midway between the unpredictability of atoms and the unpredictability of God. Atoms are small pieces of our mental apparatus, and we are small pieces of God's mental apparatus. Our minds may receive inputs equally from atoms and from God. This view of our place in the cosmos may not be true, but it is compatible with the active nature of atoms as revealed in the experiments of modern physics. I don't say that this personal theology is supported or proved by scientific evidence. I only say that it is consistent with scientific evidence.

I do not claim any ability to read God's mind. I am sure of only one thing. When we look at the glory of stars and galaxies in the sky and the glory of forests and flowers in the living world around us, it is evident that God loves diversity. The principle of maximum diversity says that the laws of nature, and the initial conditions at the beginning of time, are such as to make the universe as interesting as possible. As a result, life is possible but not too easy. Maximum diversity often leads to maximum stress. In the end we survive, but only by the skin of our teeth. This is the confession of faith of a scientific heretic. Perhaps I may claim as evidence for progress in religion the fact that we no longer burn heretics.

That is enough about me. Let me talk now about the great transformations of the world that we are facing in the future. All through our history, we have been changing the world with our technology. Our technology has been of two kinds, green and grey. Green technology is seeds and plants, gardens and vineyards and orchards, domesticated horses and cows and pigs, milk and cheese, leather and wool. Grey technology is bronze and steel, spears and guns, coal and oil and electricity, automobiles and airplanes and rockets, telephones and computers. Civilization began with green technology, with agriculture and animal-breeding, ten thousand years ago. Then, beginning about three thousand years ago, grey technology became dominant, with mining and metallurgy and machinery. For the last five hundred years, grey technology has been racing ahead and has given birth to the modern worlds of cities and factories and supermarkets.

The dominance of grey technology is now coming to an end. During the last fifty years, we have achieved a fundamental understanding of the processes occurring in living cells. With understanding comes the ability to exploit and control. Out of the knowledge acquired by modern technology, modern biotechnology is growing. The new green technology will give us the power, using only sunlight as a source of energy and air and water and soil as sources of materials, to manufacture and recycle chemicals of all kinds. Our grey technology of machines and computers will not disappear, but green technology will be moving ahead ever faster. Green technology can be cleaner, more flexible and less wasteful, than our existing chemical industries. A great variety of manufactured objects could be grown instead of made. Green technology could supply human needs with far less damage to the natural environment. And green technology could be a great equalizer, bringing wealth to the tropical areas of the world which have most of the sunshine, most of the human population, and most of the poverty.

I am saying that green technology could do all of these good things, bringing wealth to the tropics, bringing economic opportunity to the villages, narrowing the gap between rich and poor. I am not saying that green technology will do all these good things. "Could" is not the same as "will." To make these good things happen, we need not only the new technology but the political and economic conditions that will give people all over the world a chance to use it. To make these things happen, we need a powerful push from ethics. We need a consensus of public opinion around the world that the existing gross inequalities in the distribution of wealth are intolerable. In reaching such a consensus, religions must play an essential role. Neither technology alone nor religion alone is powerful enough to bring social justice to human societies, but technology and religion working together might do the job.

We all know that green technology has a dark side, just as grey technology has a dark side. Grey technology brought us hydrogen bombs as well as telephones. Green technology brought us anthrax bombs as well as antibiotics. Besides the dangers of biological weapons, green technology brings other dangers having nothing to do with weapons. The ultimate danger of green technology comes from its power to change the nature of human beings by the application of genetic engineering to human embryos. If we allow a free market in human genes, wealthy parents will be able to buy what they consider superior genes for their babies. This could cause a splitting of humanity into hereditary castes. Within a few generations, the children of rich and poor could become separate species. Humanity would then have regressed all the way back to a society of masters and slaves. No matter how strongly we believe in the virtues of a free market economy, the free market must not extend to human genes.

A few weeks ago I was attending Mass in St. Stephen's church in England. In Princeton I am Presbyterian, but in England I am Catholic because I go to Mass with my sister. The reading from the gospel of

St. Matthew told of the angry Jesus driving the merchants and money-changers out of the temple, knocking over the tables of the money-changers and spilling their coins on the floor. Jesus was not opposed to capitalism and the profit motive, so long as economic activities were carried on outside the temple. In the parable of the talents, he praises the servant who used his master's money to make a profitable investment, and condemns the servant who was too timid to invest. But he draws a clear line at the temple door. Inside the temple, the ground belongs to God and profit-making must stop.

While I was listening to the reading, I was thinking how Jesus's anger might extend to free markets in human bodies and human genes. In the time of Jesus and for many centuries afterwards, there was a free market in human bodies. The institution of slavery was based on the legal right of slave-owners to buy and sell their property in a free market. Only in the nineteenth century did the abolitionist movement, with Quakers and other religious believers in the lead, succeed in establishing the principle that the free market does not extend to human bodies. The human body is God's temple and not a commercial commodity. And now in the twenty-first century, for the sake of equity and human brotherhood, we must maintain the principle that the free market does not extend to human genes. Let us hope that we can reach a consensus on this question without fighting another civil war. Scientists and religious believers and physicians and lawyers must come together with mutual respect, to achieve a consensus and to decide where the line at the door of the temple should be drawn.

Like all the new technologies that have arisen from scientific knowledge, biotechnology is a tool that can be used either for good or for evil purposes. The role of ethics is to strengthen the good and avoid the evil. I see two tremendous goods coming from biotechnology in the next century, first the alleviation of human misery through progress in medicine, and second the transformation of the global economy through green technology spreading wealth more equitably around the world. The two great evils to be avoided are the use of biological weapons and the corruption of human nature by buying and selling genes. I see no scientific reason why we should not achieve the good and avoid the evil. The obstacles to achieving the good are political rather than technical. Unfortunately a large number of people in many countries are strongly opposed to green technology, for reasons having little to do with the real dangers. It is important to treat the opponents with respect, to pay attention to their fears, to go gently into the new world of green technology so that neither human dignity nor religious conviction is violated. If we can go gently, we have a good chance of achieving within a hundred years the goals of ecological sustainability and social justice that green technology brings within our reach.

Now I have five minutes left to give you a message to take home. The message is simple. "God forbid that we should give out a dream of our own imagination for a pattern of the world." This was said by Francis Bacon, one of the founding fathers of modern science, almost four hundred years ago. Bacon was the smartest man of his time, with the possible exception of William Shakespeare. Bacon saw clearly what science could do and what science could not do. He is saying to the philosophers and theologians of his time: look for God in the facts of nature, not in the theories of Plato and Aristotle. I am saying to modern scientists and theologians: don't imagine that our latest ideas about the Big Bang or the human genome have solved the mysteries of the universe or the mysteries of life. Here are Bacon's words again: "The subtlety of nature is greater many times over than the subtlety of the senses and understanding." In the last four hundred years, science has fulfilled many of Bacon's dreams, but it still does not come close to capturing the full subtlety of nature. To talk about the end of science is just as foolish as to talk about the end of religion. Science and religion are both still close to their beginnings, with no ends in sight. Science and religion are both destined to grow and change in the millennia that lie ahead of us, perhaps solving some old mysteries, certainly discovering new mysteries of which we yet have no inkling. After sketching his program for the scientific revolution that he foresaw, Bacon ends his account with a prayer: "Humbly we pray that this mind may be steadfast in us, and that through these our hands, and the hands of others to whom thou shalt give the same spirit, thou wilt vouchsafe to endow the human family with new mercies." That is still a good prayer for all of us as we begin the twenty-first century.

Science and religion are two windows that people look through, trying to understand the big universe outside, trying to understand why we are here. The two windows give different views, but they look out at the same universe. Both views are one-sided, neither is complete. Both leave our essential features of the real world. And both are worthy of respect.

Trouble arises when either science or religion claims universal jurisdiction, when either religious dogma or scientific dogma claims to be infallible. Religious creationists and scientific materialists are equally dogmatic and insensitive. By their arrogance they bring both science and religion into disrepute. The media exaggerate their numbers and importance. The media rarely mention the fact that the great majority of religious people belong to moderate denominations that treat science with respect, or the fact that the great majority of scientists treat religion with respect so long as religion does not claim jurisdiction over scientific questions. In the little town of Princeton where I live, we have more than twenty churches and at least one synagogue, providing different forms of worship and belief for different kinds of people. They do more than any other organizations in the town to hold the community together. Within this community of people, held together by religious traditions of human brotherhood and sharing of burdens, a smaller community of professional scientists also flourishes.

I look out from the pampered little community of Princeton, which Einstein describes in a letter to a friend in Europe as "a quaint and ceremonious village, peopled by demi-gods on stilts." I look out from this community of bankers and professors to ask, what can we do for the suffering multitudes of humanity in the world outside. The great question for our time is, how to make sure that the continuing scientific revolution brings benefits to everybody rather than widening the gap between rich and poor. To lift up poor countries, and poor people in rich countries, from poverty, to give them a chance of a decent life, technology is not enough. Technology must be guided and driven by ethics if it is to do more than provide new toys for the rich. Scientists and business leaders who care about social justice should join forces with environmental and religious organizations to give political clout to ethics. Science and religion should work together to abolish the gross inequalities that prevail in the modern world. That is my vision, and it is the same vision that inspired Francis Bacon four hundred years ago, when he prayed that through science God would "endow the human family with new mercies."

Thank you.

Appendix 3

Although this essay, "Erebus" by George Dyson, is not about Professor Dyson directly, it is reproduced here with its author's permission. I cannot resist including it for three reasons:

1. Professor Dyson sent it to us unsolicited because he felt that we should read it. We assume that it illustrates values which he held dear, containing lessons to be passed along.

2. These lessons include appreciating that responsible people with whom one disagrees hold their opinions for reasons that deserve a hearing. One can still be friends with those who are on the opposite side of a controversial issue. Dialog over sourdough pancakes is more effective than judgment.

3. This article also gives us a sample glimpse of the interesting adventures of the Dyson family.

EREBUS

by George Dyson

In Memoriam of Robert Hunter, 1941–2005

(Erebus: the son of Chaos and brother of Night... the underground cavern through which the Shades had to walk in their passage to Hades...)

Robert Hunter, high-school dropout, Winnipeg-hardened journalist, and founding father of Greenpeace was a true genius, a close neighbor, and a true friend. He was the first person to put my name in print and the last person to bail me out of jail. He wrote half a dozen books and a regular column for a succession of newspapers, on a manual typewriter, using only two fingers, one letter at a time.

I am writing this sitting in the chair in which he wrote EREBUS—his subterranean first novel, set in a Winnipeg slaughterhouse, the prehistoric darkness out of which the unbounded faith that led to Greenpeace was later brought to light. It is the least attractive but most comfortable chair I own. The arrival of Bob's first grandchild, in a one-room cabin on the beach in Belcarra Park, British Columbia, necessitated casting it adrift.

"You know, he wrote EREBUS in that chair," explained Conan Hunter as I carried the brown, five-wheeled monster off down the trail.

Bob was universally respected, by friends and foes alike. So, especially for fellow old-timers mourning the loss of "Uncle Bob," here's a glimpse from the other side of the original Greenpeace front. Exactly 28 years after the voyage of the *Phyllis Cormack* from Kitsilano to Amchitka, I spent three days in a cabin among the San Juan mountains in Colorado with a retired US Air Force general and a retired US Air Force colonel who (as I realized only after we were well into our two days of conversation) had managed the test series in the Aleutian Islands that Bob Hunter and crew had been trying to prevent.

As the first snowstorm of the season sent dark clouds down from the mountains, and the first elk hunters headed up to the meadows of Vallecito Creek, we ate huge plates of sourdough pancakes from a pungent, fiercely persistent pot of sourdough that the Colonel—a second-generation hard-rock miner, and eyewitness to over one hundred nuclear weapon tests—had begun as a physics student at Berkeley and kept going for fifty-four years, non-stop. This particular sourdough had even been known, during one especially tough winter, to consume leftover wild game, digesting it without a trace. Another winter, a visitor who concluded the pot fermenting in the back of the kitchen was something gone bad threw it out, but the Colonel contacted a neighbor who had adopted a subculture and brought the original culture back to life. How could someone who devoted such care to preserving a single colony of one-celled organisms work so hard to perfect a new species of weapons capable of destroying all life on earth? We dug into our pancakes, and I asked my hosts about Castle Bravo, the 15-megaton test conducted at Bikini in 1954. "I'll tell you about that big one," explained the Colonel. "I had seen up until that time maybe 50 shots at least, atmospheric shots out at the test site, so I wasn't really startled. I knew it was going to be big, but Malumphy and I were at least 30 miles from ground zero. So when the order came on for countdown we put on our dark goggles. And sure enough it went off and it was a full two minutes anyway before we took off our goggles and then it was so awesome that all Malumphy could say was, 'My God, my God, my God.' It was just absolutely awesome. The fireball was still out there radiating, thirty miles away, and this was two minutes after the thing had gone off." We discussed certain technical details of the Castle test series, and the personalities involved, and then the Colonel gave me a message to pass along, delivered with exactly the same passion I knew so well from listening to Bob. "I wish people could understand what would happen if one of these megatons ever got over to these cities!" he exclaimed. "I wish to hell these people could see something like that. And now we have a new generation that says 'so what.' You are going to have to keep indoctrinating the troops, indoctrinating the people as to what these things are. Or they will forget."

The Colonel (& the General) truly believed that what they were doing was right. They would have preferred to live in a world without nuclear weapons, but believed that to survive in a world with nuclear weapons required tests. They saw the Safeguard missile system (whose warhead was to be tested at Amchitka) as a non-retaliatory preventative against an intentional Soviet first strike, and as our best defense against an attack launched by insanity or by mistake. Was it crazy to be digging a 7-foot diameter hole a mile deep, backfilling with 15,000 tons of rock, and then detonating the equivalent of 5 million tons ofTNT? You bet! Only an actual test, however, would demonstrate not only whether the new W-71 design worked as advertised, but, more importantly, whether what the physicists predicted about the effects of such an intense x-ray flux on incoming warheads was true or not. The True Believers had been working nearly a decade towards this single test. They had counted on everything, but, they had not counted on Bob. As the winds gusted to 80 knots that autumn at Amchitka, they realized that anyone out there in an 80-foot purse seiner from British Columbia must be driven by beliefs even stronger than theirs.

They revealed a couple of details that I regret I never had the chance to share with Bob. First of all, as the "Don't Make a Wave" movement launched in Vancouver gained in strength, culminating in a supreme-court decision in Washington, D.C., a plan had been put in place so that if the final word came through from Washington to cancel the test (five years in on-site preparation, and with a technically irreversible instrumentation countdown already underway) there would have been a "communication failure" and the test would have gone ahead. There was nothing the *Phyllis Cormack* or *Edgewater Fortune* could have done to stop the test. You did the right thing, Bob, when you turned back.

The Cannikin test, yielding just under 5 megatons, was detonated in a 52-foot diameter cavern excavated at the bottom of a 6,000-foot deep shaft.

The last two people to finish work down there, a seismically-active mile under the surface of Amchitka Island, were two native Americans, trusted with the final task of arming the warhead, making the final connections, and getting winched up to the surface quickly since the cavern was leaking and filling up with water and there could be no going back.

Bob loved irony. It sustained him in life. So I think Bob would appreciate the irony that as he, Captain John Cormack, and their fellow Warriors of the Rainbow were heading out into the Gulf of Alaska to stop the bomb—in which they both failed and succeeded, because since that bomb (and Greenpeace) exploded, we have never conducted another such test—two native Americans 6,000 feet down, sweating away in the depths of a real Erebus, working to set the damn thing off.

George Dyson, 2 May 2005 (with apologies for unchecked facts)

Notes and References

Preface

[1] Neuenschwander, D.E. (2016).

Chapter 1. In Community

[1] Originally published in *The American Journal of Physics,* Dyson, F.J., (1991). The speech is reprinted in Ch. 16 of *From Eros to Gaia* [Dyson, F.J. (1992)].

[2] The Dyson equations occupy a central role in quantum electrodynamics, a set of coupled equations that relate to each other the expressions for vertices (interaction events) and propagators (particles moving freely).

[3] Neuenschwander, D.E. (2003).

Chapter 2. Walking with Grandfather

[1] Marshall, J.M. III (2005), p. 2.

[2] The "six faces of science" were described by Professor Dyson in his 1991 Oersted Medal speech [Dyson, F.J. (1991)]. Professor Dyson said that science is a hexagonal mountain with six faces: three beautiful and three ugly faces. The beautiful faces are (1) science as subversion of authority, (2) science as an art form, and (3) science as an international club. The three ugly faces are (4) science presented as an authoritative discipline, (5) science tied to mercenary and utilitarian ends, and (6) science tainted by its association with weapons of mass murder.

[3] Alas, Professor Dyson and Richard Feynman were not able to follow Route 66 through central Oklahoma on their "Ride to Albuquerque" [*DU* Ch. 6] because floodwaters compelled them detour to the north.

[4] Dyson, F.J. speeches: "Science and Religion," Statement to the Committee on Human Values, National Conference of Catholic Bishops, Detroit, Sept. 16, 1986. Expanded as the first chapter of *Infinite in All Directions* [Dyson, F.J. (2004)]; the preface to the Chinese edition of *DU*; Convocation Address "The Scientist as Rebel," given

at Texas Christian University, Ft. Worth, April 15, 1993; "Remarks at the Dedication of the Science and Technology Wing of Science Hall, Rider College," March 13, 1992; "Seven Ages of Man," Lecture to NTT DATA New Paradigm Session, Tokyo, August 21, 1992.

[5] More details about Ted Taylor's ice ponds are told in *Imagined Worlds* [Dyson, F.J. (1997)], pp. 40–46.

[6] See Professor Dyson's book *Weapons and Hope* [Dyson, F.J. (1984)].

[7] Dyson, F.J. (1986).

[8] Chapter 1 of *Infinite in All Directions* [Dyson, F.J. (2004)].

[9] Yates, B. (2016). The Two Windows project was led by Professor Ron Wright, SNU Dept. of Psychology, and funded by the Templeton Foundation. Freeman Dyson's quote was the inspiration behind the project name.

[10] Dyson, G.W. (1997b).

[11] By this line Garrison Keillor closed his weekly National Public Radio program, "Prairie Home Companion" whose final set was "The News from Lake Wobegone." The Lake Wobegon stories were derived from Keillor's book *Lake Wobegone Days* (Penguin Books, 1985).

[12] *DU*, "About the Author," p. 285.

Chapter 3. A Stone in Chartres Cathedral

[1] Gombrich, E.H. (1995), pp. 202–205. In this passage Gombrich describes great transitions in art that occurred during the 13th century, when "the history of great artists" began.

[2] Modified from *Song of Solomon* 8:6, King James Version.

[3] Dyson, Freeman J. (1960). "Search for Artificial Stellar Sources of Infrared Radiation," *Science* **131** (3414), 1667–1668.

[4] Albom, M. 2017, p. 133.

[5] Dyson, F.J. (1979b).

[6] Dyson, F.J. (2004).

[7] See Dyson, F.J. (1999), *Origins of Life.*

Chapter 4. Two Windows

[1] D.E. Neuenschwander, "Using *Disturbing the Universe* by Freeman Dyson as the Textbook in a 'Science, Technology, and Society' Course," *AAPT Announcer* **25** (July 1995), p. 92, Paper FC11 at the

1995 AAPT Summer Meeting at Gonzaga University, Spokane, Washington, Aug. 7–12, 1995.

[2] Freeman J. Dyson, abstract for the October, 12, 1995 colloquium of the George Washington University physics department.

[3] The Society of Physics Students, founded in 1968, is the physics society geared towards student physicists and physics appreciators. Founded in 1968 and sponsored by the American Institute of Physics, SPS offers an entrée into the physics profession with regional meetings, a peer-reviewed journal (*The Journal of Undergraduate Reports in Physics*), a quarterly magazine (*The SPS Observer*) and a variety of scholarships, awards, and internship opportunities. Sigma Pi Sigma, founded in 1921, is the physics honor society, accredited by the Association of College Honor Societies. In 1968 Sigma Pi Sigma merged with AIP upon the founding of SPS. Sigma Pi Sigma can be seen as a physics alumni association, engaging not only traditional physicists, but the so-called "hidden physicists," persons with physics degrees who are engaged in a variety of professions across the wider society. Sigma Pi Sigma hosts the quadrennial Sigma Pi Sigma Congress, publishes the biannual magazine *Radiations,* and offers a variety of awards and scholarships.

[4] George Dyson on trees thinking and boats having souls, Dyson, G. (1997a), pp. xi–x.

[5] Barbour, I. (1990).

[6] Conklin. E.G. (1925), p. 452.

[7] Freeman J. Dyson, "The Mutability of Science," a chapter in *How Large is God?* (Templeton Foundation, 1997).

[8] Chapter 5, "Ethics," of *Imagined Worlds* [Dyson, F.J. (1997)].

[9] Silver, L.M. (1998).

[10] Steinbeck, J. (1939, 1957), pp. 154–155 of the 1967 printing.

[11] Horgan, J. (1996).

[12] For more details on the missed opportunity by physicists in 1939, see *The Scientist as Rebel*, Ch. 12, "The Force of Reason" [Dyson, F.J. (2006)].

Chapter 5. Not Jonah

[1] Templeton Foundation (2006).

[2] D.N. to Wilbert Forker, letter of nomination.

[3] Dyson, F.J. (1986).
[4] Dyson, F.J. (1997).
[5] Dyson, F.J. (1979b).
[6] *Matthew* 5:9.
[7] "Dyson Family Chronicle," January 1999.
[8] Silver, L. (1998).
[9] Dyson, F.D. (1999).
[10] Diamond, J. (1999).

Chapter 6. The World Soul

[1] Panofsky, W.K.H. (1985).
[2] See also Professor Dyson's *Weapons and Hope* [Dyson, F.J. (1984)].
[3] Silver, C. (1998).
[4] Gleiser, M. (1997).
[5] Tippett, K. (2010), pp. 20–27.
[6] *On Being*, previously *Sound and Spirit*, is a weekly radio program of interviews and discussion, hosted by Christa Tippett and broadcast weekly on National Public Radio.
[7] Yates, B. (2016).
[8] Giberson, K. (1993).
[9] Karl Giberson to STS class, by letter of 11 Dec. 2000.
[10] George Will (2005).
[11] Jacob Bronowski (1973), p. 374.
[12] The poetry-reading chemistry teacher was Eric James [Schewe, P.F. (2013), p. 10]. *The Scientist as Rebel* [Dyson, F.J. (2006)] is dedicated to Eric and Cordelia James. The preface of *Rebel* offers a wonderful introduction to them. "Cordelia fought bravely for fifty years at Eric's side against the forts of folly…" "He had a Ph.D. in chemistry, but he understood that it made no sense to bore us with formal lectures about chemical reactions which we could learn about much quicker from textbooks. So he put aside ferrous and ferric oxides and read us the latest poems of Aden and Isherwood and Dylan Thomas and Cecil Day Lewis, the poets who were then speaking for the younger generation in the first desperate years of World War II," (pp. *xiii–xiv*).

Chapter 7. Not With Words, but With a Smile

[1] "The Astronomical Basis of Life on Earth" course examines astronomical habitat that is necessary for life as we know it to exist. In a nutshell: life as we know it needs real estate, the right chemicals, energy, lots of time for evolution to operate, stability during that time, and so on. For the longer story see Finkenbinder & Neuenschwander, "The Chainsaw and the White Oak: (2001).

[2] Professor Dyson's Templeton Speech, Appendix 2.

[3] See "All Night Walker Sonata" [Abley, M. (2007)].

[4] In 2013 during a breakfast conversation I asked Professor if he was "fluent in Russian," thinking he would be fluent since at the Arms Control and Disarmament Agency he studied Russian documents. He answered "No." Imme immediately explained, "His standard of fluency is very high."

[5] See Professor Dyson's contributed chapter in Chaisson & Kim (1999), pp. 53–59.

[6] Butow, R. (1954); see also Giovannitti & Freed (1965).

[7] *Luke* 10:38–42, New International Version.

[8] The lyrics of this hymn were written by John Keble (1792–1866) and published in *The Christian Year* (1827). Minor variations of lyrics exist among hymn books. I was able to find the following version, cited here in its entirety, the long version that Alice and Freeman preferred to hear softly recited by their mother as they drifted off to sleep:

> 'Tis gone, that bright and orb'ed blaze
> Fast fading from our wistful gaze;
> Yon mantling cloud has hid from sight
> The last faint pulse of quivering light.
>
> Sun of my soul, Thou Savior dear,
> It is not night if Thou be near;
> O may no earthborn cloud arise,
> To hide Thee from Thy servant's eyes.
>
> When the soft dews of kindly sleep
> My wearied eyelids gently steep,
> Be my last thought, how sweet to rest

Forever on my Savior's breast.

Abide with me from morn till eve,
For without Thee I cannot live;
Abide with me when night is nigh,
For without Thee I dare not die.

If some poor wandering child of Thine
Has spurned today the voice divine,
Now, Lord, the gracious work begin;
Let him no more lie down in sin.

Watch by the sick, enrich the poor
With blessings from Thy boundless store;
Be every mourner's sleep tonight,
Like infant's slumbers, pure and right.

Come near and bless us when we wake,
Ere through the world our way we take,
Till in the ocean of Thy love
We lose ourselves in Heaven above.

[9] *1066 and All That* is available in a recent reprinting; see Sellar and Yeatman (2010).

[10] Bronowski, J. (1973).

Chapter 8. Overcoming a Conflict Between Truth and Loyalty

[1] Friesel, Mark (2001). "Does Religion Prize Mislead Scientists?" *Physics Today* **54**, Feb. 2001, p. 82.

[2] Freeman Dyson's reply to Mark Friesel's letter, *Physics Today* **54**, Aug. 2001, p. 74.

[3] Dyson, G.B. (1997b).

[4] Wen Ho Lee, a Taiwanese-American physicist who worked at the Los Alamos National Laboratory creating computer simulations of nuclear explosions, was accused in 1999 of stealing U.S. nuclear secrets on behalf of the People's Republic of China. Lee was indicted on 59 counts by a federal grand jury. Unable to prove the accusations, the government could only charge Lee with improper handling of

restricted data, to which he pled guilty in a plea settlement. In June 2006 Lee received $1.6 million compensation from the federal government and five media companies as the outcome of a civil suit he filed against them for leaking his name to the press before formal charges had been filed against him. Lee also received an apology from Federal Judge James A. Parker for denying him bail and putting him into solitary confinement. Judge Parker criticized the government for misconduct and for misrepresenting Lee to the courts.

[5] Presumably Professor Dyson refers here to *A Short History of the World* by H.G. Wells, Macmillan, New York (1922).

[6] Schewe, P.F. (2013), p. 195.

[7] Schewe, P.F. (2013), p. 196.

[8] Bell, E.T. (1937). *Men of Mathematics: The Lives and Achievements of the Great Mathematicians from Zeno to Poincaré* (Simon & Schuster).

[9] Dyson, G.B. (1997a).

[11] Dyson, F.J. (1984), p. 120.

Chapter 9. A Problem of People's Hearts and Minds

[1] William Blake, "The Divine Image" from *Songs of Innocence* (1789).

[2] Finkenbinder *et al.* (2001).

[3] Dolly the sheep, born to her surrogate mother in July 1996, was the first mammal to be cloned from an adult cell.

[4] For more about Joseph Rotblat see Ch. 12, "The Force of Reason" in *The Scientist as Rebel* [Dyson, F.J. (2006)]. We may add another name of a person, unknown to the public but known to our campus community, who was invited to join the Manhattan Project but refused. His name was Emmett Hammer. During World War II he was a physics graduate student at the University of Kansas. When his research group at KU was asked to develop a component for the atomic bomb project, Emmett declined to participate as a matter of conscience. A decade after the war, in 1955, Professor Hammer founded our university's physics department. This part of Professor Hammer's life was told to D.N. by Polly Hammer, Dr. Hammer's daughter, in a private conversation during one of her visits to SNU.

[5] Pirsig, R.M. (1974).

[6] Pirsig describes Chris' murder in an Afterword to the 1999 edition of *ZAMM*. Pirsig wrote, "I go on living, more from force of habit than anything else." Now Robert has also passed on (d. April 24, 2017).

[7] Robert Pirsig published a second book, *Lila: An Inquiry into Morals* (Bantam Books, 1991).

Chapter 10. Not as Bad as Imposing Rules on Neighbors

[1] Albert Einstein, *Ideas and Opinions* (Crown Publishers, 1982), p. 199.

[2] Richard Rhodes (2007), *Arsenals of Folly*. The zealots mentioned explicitly by Rhodes include, among others, Paul Nitze (Ch. 6), Donald Rumsfeld and Richard Cheney (Ch. 7): "Nitze, according to a scholar of the report [NSC-68: United States Objectives and Programs for National Security]... 'wanted to sacrifice a degree of rationality in the analysis of NSC-68 in order to exaggerate the [Soviet] threat'... which is to say, the bludgeon Nitze chose to use was threat inflation. Although Nitze and his staff consulted no experts on the Soviet Union, and were neither experts themselves nor even fluent in Russian, NSC-68 is weighted with rhetorical absolutes." (p. 116). "Cheney was a serious student of political power and derived both his employment and his enjoyment from it. Whenever his private ideology was exposed, he appeared somewhat to the right of Ford, Rumsfeld, or, for that matter, Genghis Kahn." (p. 118). "Rumsfeld and Cheney... [were] opposed to the policy of détente with the Soviet Union, and they operated by stealthy internal maneuver" (p. 119).

[3] Professor Dyson refers to his review of *Isaac Newton* by James Gleick (Pantheon Books, New York, 2003). The review was published in *New York Review of Books,* July 3, 2003.

[4] After 2007, another assigned STS reading to accompany "Clades and Clones" was "All-Night Walker Sonata" by Mark Abley [Ably, M. (2007)], pp. 18–20, who writes, "...soon a language has turned into the property of elders, who mourn its withering but may be helpless to do anything about it. Such is the template of world grief."

[5] Weinberg, S. (1977), p. 154.

[6] Freeman Dyson, "On the Radiation Theories of Tomonaga, Schwinger, and Feynman," *Physical Review* **75**, Feb. 1, 1949, pp. 486–502; "The *S* Matrix in Quantum Electrodynamics, *Phys. Rev.*

75, June 1, 1949, pp. 1736–1755. For a detailed history see Schweber (1994).

[7] D.E.N. to Professor Dyson, letter of 13 November 2003.

[8] Frankl, V.E. (1963).

[9] Schweber, S.S. (1994), p. 492.

[10] Neuenschwander, D.E. (2003), "Rattlesnake University" an essay motivated by the silence of the Great Basin. The "Doubting Thomas" talk was presented in our university convocation on August 30, 2000. Thomas is one of my biblical heroes because he was honest about expressing his doubt, as told in *The Gospel of John* 20: 24–29.

[11] Pagels, E. (2003).

[12] Neuenschwander, D.E. (2000).

[13] D.E.N.'s funeral eulogy for Evonne Neuenschwander, Jan. 17, 2003.

Chapter 11. The Mailman is More Important

[1] Crawford, M.B. (2010), pp. 9–10.

[2] D.N.'s father, while a Navy recruit stationed in San Diego in 1948–1950, took a Navy course, "US Navy IC Intercommunication Course: A Study of Electricity." I don't know if it's so, but I like to think that he may have taken the course in the Barnard School. At least he would have seen the "little red schoolhouse" during his time there.

Chapter 12. The Family Next Door

[1] Sherry Turkle, *Alone Together* [Turkle, S. (2011)], p. 1.

[2] Auel, J. (1980) *The Clan of the Cave Bear* (Crown).

[3] Professor Dyson's referenced papers in Misner *et al.* (1973) include: from 1954, *Advanced Quantum Mechanics*, lithographed lecture notes, Physics Department, Cornell University (referenced in MTW Box 17.2 part 6), also available in 2nd edition (2011) from World Scientific; from 1967, "Time variation of the charge of the proton," *Phys. Rev. Lett.* **19**, 1291–1293 (MTW Sec. 37.3 and Fig. 37.2); from 1969, "Seismic response of the Earth to a gravitational wave in the 1-Hz band," *Astrophys. J.* **156**, 529-540 (MTW Sec. 37.3, Fig. 37.2); from 1972, "The fundamental constants and their time variation" in Salam and Wigner (1972), MTW Sec. 38.6.

[4] Teller, E. and Shoolery, J. (2001) *Memoirs: A Twentieth-century Journey in Science and Politics* (Perseus). Professor Dyson's review was

published in *Am. J. Phys.* **70** (2002), p. 462. It also forms Chapter 15 of *The Scientist as Rebel* [Dyson, F.J. (2006)].

[5] Dyson, F.J. (1991).

[6] Dyson, G.B. (2012).

[7] Wells, H.G., *Tono-Bungay* (Macmillan, 1909); *The Island of Doctor Moreau* (Heinemann, 1896).

[8] Huxley, J. (1932). *Brave New World* (Chatt and Windus).

[9] A linkage between Martin Luther King Jr. and Professor Dyson is made in "The Physicist, the Prophet, and Activism," Neuenschwander (2020).

[10] See Turkle, S. (2011) *Alone Together.*

[11] Dyson, F.J. (2004), "The Domestication of Biotechnology," *Chautauqua Daily*, August 24, 2004.

[12] The cartoon appears in *Dear Professor Dyson* on p. 32.

[13] Robert Pirsig, *Zen and the Art of Motorcycle Maintenance* (1974), pp. 193–199.

Chapter 13. The Varieties of Human Experience

[1] William James, *The Varieties of Religious Experience* [James, W. (1902)].

[2] Freeman Dyson, "The Varieties of Human Experience," Witherspoon Lecture, Princeton, November 6, 2003.

[3] James, W. (1902).

[4] Generalizing the principle of complementary wave-particle duality, in a 1939 essay Niels Bohr describes how there are two kinds of truth, Simple Truth and Deep Truth. The opposite of a Simple Truth is false, but the opposite of a Deep Truth is also true. See Bohr's essay, p. 240, "Discussions with Einstein" in Schlipp (1970). This statement by Bohr about Simple and Deep Truths forms a recurring theme in our STS class.

[5] Carl Sagan (1990), "Guest Comment: Preserving and cherishing the earth: an appeal for joint commitment in science and religion," *American Journal of Physics* **58** (7), July 1990, pp. 615–617. Signers of this manifesto include Professor Dyson and Hans Bethe.

[6] Pagels, E. (1979).

[7] A line from Professor Dyson's review of *The Road From Los Alamos* by Hans Bethe and Hans C. von Baeyer (Springer, 1991). The review was published in *American Journal of Physics* **60** (1), Jan. 1992, p. 91.

[8] Else, J. (1980).

[9] An interesting insight to the famous Great Books program, its genesis and its demise at the University of Chicago, may be found in Ashmore, Harry S. (1989) *Unseasonable Truths: The Life of Richard Maynard Hutchins* (Little, Brown, Co.), Chapter Ten, "The Great Conversation." A glimpse of the controversy about the Great Books curriculum at the University of Chicago appears in Robert Pirsig's *Zen and the Art of Motorcycle Maintenance* [Pirsig, R. (1974)] Pirsig was a graduate student in a course taught by Robert McKeon, one of the principals in the original Great Books program. The Great Books curriculum survives today at St. John's College in Annapolis, Maryland.

[10] The impressive role of Emma Epps as a community leader in Princeton is described by Jerome & Taylor (2005), pp. 32, 40–41.

[11] Jerome & Taylor (2005), Ch. 3.

[12] Peter Sankey (1923–1944), commissioned in the Corps of Royal Engineers (Nov. 1942), Intelligence Officer, HQ Royal Engineers, 1st Airborne Division, killed in action an Amhem. See https://www.unithistories.com/officers/1AirbDiv_officersS.htm

Chapter 14. Talk to Your Enemies

[1] Gorbachev, M. (1987), p. 215.

[2] Dyson, F.J. (1986).

[3] The first of the Ten Commandments says "Thou shalt have no other gods before Me," *Exodus* 20:3.

[4] Dyson F.J. (1984).

[5] Professor Dyson refers to the Cuban Missile Crisis.

[6] Hayes, P. and Tannenwald, N. (2003).

[7] President Eisenhower's "military-industrial complex" speech can be found at the National Archives website: https://www.archives.gov/milestone-documents/president-dwight-d-eisenhowers-farewell-address

[8] Professor Dyson refers to the bombing of the Alfred P. Murrah federal building in downtown Oklahoma City that occurred on April 19, 1995.

[9] In both documents, *DU* and Dyson, F.J. (1979b), Professor Dyson quotes Steven Weinberg's famous line from *The First Three Minutes*:

"The more the universe seems comprehensible, the more it also seems pointless." [Weinberg, S. (1977)] To this Professor Dyson replies, "I have seen the galaxies pass before me, but the Lord was not in the galaxies..." [*DU* p. 259]. Professor Dyson here adapts a passage from I Kings 19:11–12, King James Version: "[11] And he said, Go forth, and stand upon the mount before the Lord. And, behold, the Lord passed by, and a great and strong wind rent the mountains, and brake in pieces the rocks before the Lord; but the Lord was not in the wind: and after the wind an earthquake; but the Lord was not in the earthquake:[12] And after the earthquake a fire; but the Lord was not in the fire: and after the fire a still small voice."

[10] Weinberg, S. (1992).

[11] Choi-D.N. correspondence, e-mail letters of 21 March 2007.

Chapter 15. God Has a Sense of Humor

[1] Comment by ornithologist and environmental biologist Leo Finkenbinder, while birdwatching near Arenal Volcano, Costa Rica.

[2] See ASU | Universidad Latina de Costa Rica (ulatina.ac.cr)

[3] Freeman Dyson played violin in school performances at Winchester College [Schewe, P.F. (2013)], p. 11.

[4] The Patriot Act of 2001 (officially, "Uniting and Strengthening America by Providing Appropriate Tools Required to Intercept and Obstruct Terrorism (USA PATRIOT) Act of 2001") gave law enforcement agencies enhanced powers to investigate suspected terrorists, indict them and bring them to justice. Penalties for supporting and committing terrorist acts were increased. However, Section 215 of the Patriot Act violates the Fourth Amendment of the US Constitution, which requires the government to obtain a warrant and show probable cause before a person's private property can be searched.

[5] Mayer, J. (2009). The quoted phrase is in the book's subtitle.

[6] The General Lord quote appears in Weiner, T. (2005).

[7] Nina Tannenwald (2007).

[8] Rhodes, R. (1995), *Dark Sun,* Ch. 20, "Gung-Ho for the Super." Excerpts: "At the same time, Truman cut off all public and most private debate by restricting discussion of the question to the Special Committee and its staff" (p. 404). "When the Special Committee came

in [to the White House] on January 31 [1950] to recommend to him that the nation proceed with the Super, Lilienthal was prepared to argue at length that the policy was not wise. Truman cut short the discussion. 'What the hell are we waiting for?' he remembered telling them. 'Let's get on with it" (p. 407). See also Hans Bethe (1950), "The Hydrogen Bomb: II," *Scientific American* **182**, April 1950, pp. 18–23. The federal government raided the Scientific American offices, destroyed three thousand printed copies, melted down the type and confiscated every galley proof. The government did not get all copies. The article can also be found on
https://www.documentcloud.org/documents/5677508-LostSciAm#document/p1/a474212s?

Chapter 16. Family First, Friends Second, Work Third

[1] Luther Standing Bear (1933, 1978), p. 91.
[2] Neuenschwander, D.E. (2020) "The Prophet and the Physicist."
[3] Christopher Sykes documentary, *Richard Feynman: Last Journey of a Genius* (BBC, NOVA, 1988).
[4] Leighton, R. (1991).
[5] Dyson, G.B. (2012).
[6] *Star Trek* episode with Dyson Sphere:
http://www.youtube.com/watch?v=ECLvFLkvY7Y
[7] For an array of independent particles arranged in a spherical shell, each particle would orbit the star in a plane that passes through the star's center. For a solid spherical shell, portions of the shell located at latitudes above and below the star's equatorial plane would experience shear forces, tending to break the spherical shell apart. To illustrate, think of the Sun's surface as a spherical shell centered on the Sun's center. This surface exhibits differential rotation; that is, the material in the Sun's equator orbits with a greater speed than material on this surface that lies above or below the equator. Professor Dyson did not envision a rigid spherical shell, but a huge array of satellites independently orbiting the star.
[8] Dyson, Freeman J. (1960).
[9] Kaufmann, W. (1961), p. 2.
[10] D.E.N., speech for Senior Farewell Chapel, Southern Nazarene University, May 4, 2010 and May 1, 2012. Quoted song lyrics included

"The Morning" by the Moody Blues, from *Days of Future Passed* (Deram, 1967) and "Time" by Pink Floyd (EMI, from *The Dark Side of the Moon*). The speech emphasized how fast time will go by for the students after they graduate. From "The Morning:"

"Time seems to stand quite still,
In a child's world it always will…
Yesterday's dreams are tomorrow's sighs,
Watch children playing, they seem so wise…"

From "Time,"

"And then one day you find
Ten years have got behind you
No one told you when to run
You missed the starting gun.
You run and you run to catch up with the sun but it's sinking
Racing around to come up behind you again
The sun is the same in a relative way but you're older
Shorter of breath, and one day closer to death…"

[11] Eisenhower's presidential farewell speech of 1961, Ref. 7 of Ch. 14.

[12] From *Weapons and Hope:* "Throughout the history of warfare, two styles of professional soldering have alternatively prevailed, the style of brains and quick reaction and the style of stupor and mass destruction. World War I was the classic example of a war of stupor and mass destruction.... As a rule, professional soldiers who take pride in their professions prefer the weaponry of brains and quick reaction to the weaponry of mass destruction" [Dyson, F.J. (1984), pp. 162–163].

Chapter 17. Living Through Four Revolutions

[1] Excerpt of lyrics from "Revolution" by the Beatles (EMI/Apple, 1968).

[2] The Sigma Pi Sigma physics honor society was founded in 1923 at Davidson College in North Carolina. Today it is sponsored and underwritten by the American Institute of Physics, a consortium of ten physics professional societies. Sigma Pi Sigma hosts a convention, the Sigma Pi Sigma Congress, every four years (recently changed to every three years).

[3] Professor Dyson's comments on kamikaze pilots of WW2 can be also found in Ch. 29, "Religion from the Outside," of *The Scientist as Rebel*

[Dyson, F.J. (2006]. There he describes *Kamikaze Diaries: Reflections of Japanese Student Soldiers* by Emiko Ohnuki-Tierney (Univ. of Chicago Press, 2006). Here is a passage from Professor Dyson's review (pp. 351–353):

> *The diaries give us firsthand testimony of the thought and feelings of these young soldiers who knew that they were fated to die. Their thoughts and feelings are astonishingly lucid and free from illusions. Some of them expressed their feelings in poetry. All of them were hightly educated... Only one of them, Hayashi Ichizo, was religious...His Christian faith did not make self-sacrifice easier for him than for the others....*
>
> *All of the young men, including Hayashi, had a profoundly tragic view of life, mitigated only by happy memories of childhood with family and friends. They were as far as it was possible to be from the brainwashed zombies that contemporary Americans imagined them to be piloting their kamikaze planes. They were thoughtful and sensitive young men, neither religious nor nationalist fanatics.*
>
> *In his last letter to his parents, a week before his death, Nako Takanori wrote..."Working as one heart, we will plunge into the enemy vessel..."*
>
> *We have no firsthand testimony from the young men who carried out the September 11 attacks. They were not as highly educated and as thoughtful as the kamikaze pilots, and they were more influenced by religion. But there is strong evidence that they were not brainwashed zombies. They were soldiers enlisted in a secret brotherhood that gave meaning and purpose to their lives, working together in a brilliantly executed operation against the strongest power in the world...they were motivated like the kamikaze pilots, more by loyalty to their comrades than by hatred of the enemy. Once the operation had been conceived and ordered, it would have been unthinkable and shameful to not carry it out.*
>
> *Even after recognizing the great differences between the circumstances of 1945 and 2001, I believe that the kamikaze diaries give us our best insight into the state of mind of the young men who caused us such grievous harm in 2001. If we wish to understand the phenomenon of terrorism in the modern world, and*

if we wish to take effective measures to lessen its attraction to idealistic young people, the first and most necessary step is to understand our enemies. We must give respect to our enemies, as courageous and capable soldiers enlisted in an evil cause, before we can understand them. The kamikaze diaries give us a basis on which to build both respect and understanding.

[4] 2012 Congress brochure, 2012-PhysCon-Program.pdf (spsnational.org)

[5] *Gospel of John,* Ch 18.

[6] Here Professor Dyson refers to *An Inconvenient Truth* (2006), a documentary and book on climate change.

Chapter 18. The Water Meadows

[1] *Ahead of All Parting: The Selected Poetry and Prose of Ranier Maria Rilke,* translated and edited by Stephen Mitchell (Modern Library, NY, 1995), p. 191.

[2] Freeman Dyson first showed that Richard Feynman's diagrammatic method of solving QED was equivalent to the established formal methods of Julian Schwinger and Sin-Itiro Tomonaga. Feynman, Schwinger, and Tomonaga shared the 1965 Nobel Prize in Physics. The Nobel Prize rules allow a maximum of three people to share a prize in the same year in the same discipline. But Professor Dyson cheerfully makes light of it, saying it's better for people to ask you why you did not win the Nobel Prize, than for them to ask you why you did. [Schewe, P.F. (2013), p. 180]. Everyone who knows anything about QED realizes that Professor Dyson is in the equivalence class of Nobel Laureates.

[3] Prizes that have been bestowed on Professor Dyson include: American Institute of Physics and American Physical Society Dannie Heineman Prize for Mathematical Physics (1965), German Physical Society Max Planck Medal (1969), Harvey Prize (1977), Wolf Prize in Physics (1981), American Association of Physics Teachers Oersted Medal (1991), Enrico Fermi Award (1995), National Space Society's Robert Heinlein Memorial Award (2018), Templeton Prize (2000), Henri Poincaré Prize (2012). Memberships include the National Academy of Sciences, the Royal Society, and the Russian Academy of Sciences. Apologies for those I missed.

[4] Phua *et al.* (2014), p. ix.

[5] Dyson, F.J. in Phua *et al.* (2014), pp. 1, 12–13.

[6] January 2013 Dyson Family Chronicle.

[7] D.E. Neuenschwander, in Phua *et al.* (2013), pp. 308–327.

[8] In "The Blood of a Poet," Ch. 4 of *Disturbing the Universe*, Professor Dyson tells the story of Frank Thompson, an older student at Winchester College when Freeman was a young student there. Frank joined the partisans in Bulgaria during World War 2. He and his men were captured and condemned by the Nazis. As he and his comrades were led away, with Frank in the lead, they gave a clenched-fist salute. Frank Thompson was declared a national hero of Bulgaria. The railway station of the town of Prokopnik, "where the partisans fought one of their fiercest battles, was renamed Major Thompson Station." The Bulgarian gentleman who spoke with Professor Dyson said that, until he came across this passage in *Disturbing the Universe*, he always wondered why a Bulgarian railway station bore an Englishman's name. The Bulgarian gentleman thanked Professor Dyson for providing the backstory in *DU*.

[9] *Sunrayce* is an intercollegiate solar-powered car race sponsored by General Motors, the US Department of Energy, and Electronic Data Systems. Powered only by photoelectric cells that charge batteries to run electric motors, the cars are driven coast-to-coast in a multi-day race. Speeds average about 40 mph but can reach 70 mph in these lightweight, low-profile cars. While they carry so far only the driver, we recall that the performance and carrying capacity of the earliest gasoline-powered cars (c. 1900) were comparable to today's riding lawn mowers. The Sunrayce cars are proof-of-concept demonstrations that solar-powered vehicles are possible.

[10] This point is dramatically made in Ch. 1, "Stories," in *Imagined World* [Dyson, F.J. (1997)], with stories that include, among others, the development of the airplane, in contrast to the development of the airship. Today we fly around the world in airplanes (e.g., the Boeing 747) instead of airships (like the *Hindenberg*) because the development of airplanes was an evolutionary process where bad ideas were allowed to fail, while the development of airships was driven by politics and ideology, which resulted in unrealistic timelines and disasters.

[11] The Drake Equation offers a probabilistic calculation for the number of communicative civilizations per galaxy. One takes the number of stars per galaxy and makes cuts: What fraction of those stars have planets within the star's habitable zone; on what fraction of such planets does life come into existence; on what fraction of planets with life does life evolve to intelligence; for what fraction of the host star's lifetime does intelligent, communicative life survive, and so on. The only factors that can be estimated with reasonable confidence are number of stars per galaxy and perhaps the number of planets per star in a star's habitable zone, the latter due to the Kepler satellite that discovered at least one planet orbiting every star it observed. The remaining factors are completely unknown. Depending on whether one makes pessimistic or optimistic guesses of those remaining factors, the number of communicative civilizations can range from, say, one per ten thousand galaxies to hundreds of thousands per galaxy. The Drake equation, while sound in principle, is presently useless for finding real answers because most of the factors going into it are unknown.

[12] Dyson, F.J. (1999), *Origins of Life*.

Chapter 19. Society and Sanity

[1] Pirsig, R.M. (1999), *Zen and the Art of Motorcycle Maintenance*, p. 350.

[2] Dyson, F.J. (1991).

[3] Dyson, F.J. (2006), p. 28.

[4] Dyson, F.J. (1984), p. 188.

[5] Russian culture and history, so we read, owes much to Kiev in addition to Moscow and St. Petersburg. See, for example, Melvin C. Wren (1979), *The Course of Russian History* (Macmillan, NY, 1979). Wren writes in Ch. 3, "Kievan Society," that "Russians have always felt a sentimental attachment for the history of Kiev, 'the mother of Russian cities.'" (p. 41). Putin's unprovoked attack of Ukraine, with its deliberate bombardment and sieges of the civilian populations, seems to be less about a "sentimental attachment for the history of Kiev" but more about Putin's egocentric dream of restoring the Soviet empire with, of course, himself at its head.

[6] Dyson, F.J. (1986).

[7] Carr, N. (2014).

[8] Wiener, N. (1950). Professor Dyson evidently refers to another printing; the quote is found in p. 162 of the 1954 reprint of Wiener's 1950 book cited in the bibliography.

[9] Tannenwald, Nina (2007).

[10] Stockpile Stewardship replaces nuclear weapons tests with computer simulations.

[11] Thoreau, H.D. (1960), p. 56.

[12] Marshall, J.M. (2005).

[13] Neuenschwander (2016), pp. 246–247.

Chapter 20. Social Justice as Necessary for a Healthy Society

[1] Fritz Mauthner (1849–1923); from his book of modern fables, *Aus dem Märchenbuch der Wahrheit* (*Fairytale Book of Truth*), 1899.

[2] Oliver Sacks was a professor of neurology at the New York University School of Medicine, and a best-selling author. His books include *Awakenings; Hallucinations; Musicophilia: Tales of Music and the Brain; The Man Who Mistook His Wife for a Hat;* and his autobiography *On the Move* whose cover shows Sacks aboard a motorcycle.

[3] This was Senator Jim Inhofe, (R)-Oklahoma. Our letter to him was dated May 5, 2015 and signed by all class members.

[4] Dyson, F.J. (2015).

[5] The 2015 United Nations Climate Change Conference was held in Paris, November 30–December 12. The outcome was the "Paris Agreement," a consensus of 196 nations on terms directed towards reducing climate change.

[6] The Kyoto Protocol was an international treaty adopted in December 1992 in Kyoto, Japan, which committed signatories to reduce greenhouse gas emissions. The treaty went into force in February 2005.

[7] Asano, Taro (August 1970). "Theorems on the Partition Functions of the Heisenberg Ferromagnets," *Journal of the Physical Society of Japan*. **29** (2): 350–359.

[8] I know something of the guilt which Professor Dyson felt…

…Three ten-year-old boys, up in a tree, checking it out for a possible tree house. When climbing back down, one boy loses his grip and falls. His head strikes a steel barrel that happens to be standing by the tree trunk. The fallen boy does not move. The bleeding does not stop. Firemen and an ambulance appear. Their hands work feverishly over the fallen boy. Unconscious, he is whisked away. Two weeks later, never regaining consciousness, he dies...

I must live with the fact that the idea to climb that tree was mine. Although I know that fifth-grade boys climb trees every day, and although I did not place the barrel there or jostle a branch to cause the fall, I also know that, had I not suggested the tree house project, Charles Spicer could have been someone's grandpa today. Like Professor Dyson with Taro Asano, a sense of guilt will always remain with me. This is not a book about Charles Spicer. But this is my chance to publicly honor his memory. –D.E.N.

[9] We were struck by the similarities between the 2016 presidential campaign and the subsequent administration of Donald Trump on the one hand, and those of Benito Mussolini and Adolf Hitler on the other. Even before Trump was elected in 2016, while he was seeking the nomination, observers in our class identified him as a Fascist. See William L. Shirer (1961), *The Rise and Fall of the Third Reich: A History of Nazi Germany* Chs. 1–6 (Secher and Warburg, London); Albert Speer, *Inside the Third Reich,* Chs. 1–3, tr. By Richard and Clara Winston (Macmillan, 1970); William L. Shirer, *Berlin Diary: The Journal of a Foreign Correspondent 1934–1941* (Ryerston Press 1941, Penguin Books 1979). See also Timothy Snyder, *On Tyranny* (Tim Duggan Books, NY, 2017). Fascism is not a set of responsible principles that guide policy decisions towards the good of the population. Fascism is a strategy for seizing and holding power. Any means, however brutal, are assumed to be justified by the ends. The basic features of such means include the following:

1. Promote extreme nationalism, making it a secular religion.
2. Tell lies continuously and so brazenly that they are eventually accepted as truth by the masses. Demonize anyone who does not agree with you.
3. Find a minority group or groups to be the victims of "othering," who are made prime targets and unjustly blamed for the nation's troubles. Nurture "us vs. them" attitudes. Preach to your base of followers that *they* are victims of the "other."
4. Having stoked a fire of widespread fear and resentment, proclaim to the nation that "I am the only one who can fix it."

5. Once in power, implement cruelty in policies.

[10] Dyson Family Chronicle for 2016.

Chapter 21. The Serenity of Old Age

[1] Excerpt from "Silence" by Edgar Lee Masters (1915).

[2] Severt Young Bear confirms this from personal experience: "As we are sitting here talking late into the day, it's important to remember that although we're considering making all this into a book, everything we bring up is part of the oral tradition, the *Lakol wichoh'an* (Lakota heritage). That spoken word and the memory that catches and keeps it are at the center of our tradition. When I was younger, I once asked my dad to tape-record something for me because I wanted to be able to remember it. He refused and said: 'Son, I will tell you all about it, but I don't want you to record it. If it's important enough to you, you will and must remember it in your mind. Concentrate and you'll remember what you're told and it will stay with you. If you record it on a machine, you'll lose it." Severt Young Bear and R.D. Theisz, *Standing in the Light: A Lakota Way of Seeing* (University of Nebraska Press, 1994).

[3] Familiar programs of service that are available nationally to all students are post-graduate assignments such as the Peace Corps, Teach for America, etc.

[4] https://www.ipcc.ch/assessment-report/ar5/

[5] Another person with whom the STS class has corresponded is Kenneth Ford. His article, "Working (and Not Working) on Weapons" [Ford, K. (2005)] is assigned reading. See also *Building the H Bomb* [Ford, K.W. (2015)].

Chapter 22. Listening to Almustafa

[1] Gibran, K. (1923).

[2] Dyson, F.J. and Neuenschwander, D.E. (2019).

Chapter 23. Doubt, Faith, and Peaceful Coexistence

[1] M. King Hubbert, "Nuclear Energy and the Fossil Fuels." Presented before the Spring Meeting of the Southern District, American Petroleum Institute, San Antonio, Texas, March 7–8–9, 1956.

[2] Bertrand Russell (1957), p. vi.

[3] Walter Kaufmann (1961), p. 2.
[4] *The Civil War: A Film by Ken Burns* (1990). Florentine Films and WETA-TV.
[5] Theodore White (1978), pp. 304–305, 511.

Appendix 1

[1] Butler, O. (1993) *The Parable of the Sower* (Four Walls Eight Windows).

Bibliography

Abley, Mark (2007). "All-Night Walker Sonata," *World Literature Today*, Sept.–Oct. 2007.

Albom, Mitch (1997, 2017). *Tuesdays with Morrie* (Broadway Books).

Auel, Jean M. (1980). *The Clan of the Cave Bear* (Crown).

Bell, E.T. (1937, 1965). *Men of Mathematics: The Lives and Achievements of the Great Mathematicians from Zeno to Poincaré* (Simon & Schuster).

Barbour, I.G. (1990). *Religion in an Age of Science*, Vol. 1 (Harper Collins).

Bronowski, Jacob (1973). *The Ascent of Man* (Little, Brown, & Co.).

Butow, Robert (1954). *Japan's Decision to Surrender* (Stanford University Press).

Carr, Nicholas (2014). *The Glass Cage: Automation and Us* (W.W. Norton & Co.).

Chaisson, Eric J. and Kim, Tae-Chang (1999). *The 13 Labor: Improving Science Education* (Gordon & Breach). Chapter 6 is Professor Dyson's essay, "Tolstoy, Napoleon, and Gompers."

Conklin, E.G. (1925). "Science and Faith of the Modern," *Scribner's Magazine* **78**, p. 452.

Crawford, M.B. (2010). *Shop Class as Soulcraft: An Inquiry into the Value of Work* (Penguin).

Diamond, Jared (1999). *Guns, Germs, and Steel: The Fates of Human Societies* (Norton).

Dyson, F.J. (1960). "Search for Artificial Stellar Sources of Infrared Radiation," *Science* **131** (3414), 1667–1668.

Dyson, F.J. (1979a). *Disturbing the Universe* (Basic Books).

Dyson, F.J. (1979b). "Time without end: Physics and biology in an open universe," *Review of Modern Physics* **51** (3), 447.

Dyson, F.J. (1984). *Weapons and Hope* (Harper Colophon Books).

Dyson, F.J. (1986). "Science and Religion," Statement to the Committee on Human Values, National Conference of Catholic Bishops, delivered in Detroit, Michigan, September 16, 1986 (speech text, private communication). This speech was expanded into Ch. 1 of *Infinite in All Directions*.

Dyson, F.J. (1991). "To Teach or Not to Teach," Freeman J. Dyson's acceptance speech for the 1991 Oersted Medal presented by the American Association of Physics Teachers, 22 January 1991, *American Journal of Physics* **59** (6), 491–495.

Dyson, F.J. (1992). *From Eros to Gaia* (Penguin Books).

Dyson, F.J. (1997). *Imagined Worlds* (Harvard University Press).

Dyson, F.J. (1999). *Origins of Life* (Cambridge Univ. Press).

Dyson, F.J. (1999). *The Sun, the Genome, and the Internet: The Tools of Scientific Revolutions* (Oxford University Press).

Dyson, F.J. (2004). *Infinite in All Directions* (Harper Perennial).

Dyson, F.J. (2004). "The Domestication of Biotechnology," *Chautauqua Daily*, August 24, 2004.

Dyson, F.J. (2006). *The Scientist as Rebel* (New York Review of Books).

Dyson, F.J. and Neuenschwander, D.N. (2019). "Youth Engaging Almustafa: Cross-Generational Interactions in 'Science, Technology, and Society' Education," *The Physics Educator* **1** (2), https://doi.org/10.1142/S2681339519200014 https://www.worldscientific.com/worldscinet/tpe.

Dyson, F.J. (2015). *Dreams of Earth and Sky* (New York Review of Books).

Dyson, G.B. (1997a). *Darwin Among the Machines: The Evolution of Global Intelligence* (Helix Books).

Dyson, G.B. (1997b). *Baidarka: The Kayak* (Alaska Northwest Books).

Dyson, G.B. (2002). *Project Orion: The True Story of the Atomic Spaceship* (Owl Books).

Dyson, G.B. (2012). *Turing's Cathedral: The Origins of the Digital Universe* (Vintage Books).

Else, Jon (1980). *The Day After Trinity: J. Robert Oppenheimer and the Atomic Bomb* (DVD documentary, Pyramid).

Finkenbinder, Leo R. and Neuenschwander, D.E. (2001). "The Chainsaw and the White Oak: From Astrobiology to Environmental Sustainability," *Radiations* **7**, Spring 2001, pp. 5–11.

Ford, Kenneth W. (2005). "Working (and Not Working) on Weapons," *Radiations* **11**, Spring 2005, pp. 5–7.

Ford, Kenneth W. (2015). *Building the H Bomb: A Personal History* (World Scientific, Singapore).

Frankl, Victor E. (1963, 1970). *Man's Search for Meaning: An Introduction to Logotherapy* (Washington Square Press).

Giberson, Karl (1993). *The Unholy War: The Conflict Between Science and Religion* (Beacon Hill Press, Kansas City).

Gibran, Kahlil (1923). *The Prophet* (Knopf 1923, my copy the 71st printing (1964).

Giovannitti, L. and Freed, F. (1965). *The Decision to Drop the Bomb: A Political History* (Coward-McMann).

Gleiser, Marcelo (1997). *From Creation Myths to the Big Bang* (Dartmouth College Press, Lebanon, NH).

Gombrich, E.H. (1995). *The Story of Art*, 15th Ed. (Phaidon Press).

Gorbachev, Mikhail (1987). *Perestroika: New Thinking for Our Country and the World* (Harper & Row, New York).

Hayes, P. and Tannenwald, N. (2003). "Nixing Nukes in Vietnam" *Bulletin of Atomic Scientists* **59**, pp. 52–59.

Horgan, J. (1996). *The End of Science: Facing the Limits of Knowledge in the Twilight of the Scientific Age* (Helix Books).

Kaufmann, Walter (1961). *The Faith of a Heretic* (Anchor Books, New York).

Leighton, Ralph (1991). *Tuva or Bust! Richard Feynman's Last Journey* (W.W. Norton, New York).

James, William (1902, 2008). *The Varieties of Religious Experience: A Study In Human Nature* (Manor, 2008 printing).

Jerome, Fred and Taylor, Rodger (2005). *Einstein on Race and Racism* (Rutgers University Press).

Luther Standing Bear (1933, 1978). *Land of the Spotted Eagle* (Univ. of Nebraska Press).

Kaufmann, Walter (1961). *The Faith of a Heretic* (McGraw Hill).

Marshall, Joseph M. III (2005). *Walking with Grandfather* (Sounds True, Boulder, CO).

Mayer, Jane (2009). *The Dark Side: The Inside Story of How the War on Terror Turned Into a War on American Ideals* (Anchor Books).

Neuenschwander, D.E. (1995). "Using *Disturbing the Universe* by Freeman Dyson as the Textbook in a 'Science, Technology, and Society' Course," presented at the Summer 1995 AAPT meeting, Gonzaga University, Spokane, WA, Aug. 11, 1995. *AAPT Announcer* **25** (July 1995), 92. The abstract mentions how "student's respect for the text has been an essential component of the course's success in accomplishing its mission."

Neuenschwander, D.E. (2003). "Rattlesnake University" (*SPS Observer*, Winter 2003).

Neuenschwander, D.E. (2000). "Conversations with Ghosts," *Am. J. Phys.* **69** (3), March 2000, pp. 251–254.

Neuenschwander, D.E., Ed. (2016). *Dear Professor Dyson: Twenty Years of Correspondence Between Freeman Dyson and Undergraduate Students on Science, Technology, Society and Life* (World Scientific).

Neuenschwander, D.E. (2020). "The Prophet, the Physicist, and Activism" *Radiations* (Fall 2020), pp. 19–21.

O'Neil, Cathy (2016). *Weapons of Math Destruction: How Big Data Increases Inequality and Threatens Democracy* (Crown, New York).

Pagels, Elaine (2003). *Beyond Belief* (Random House).

Panofsky, Wolfgang K.H. (1985). "The Strategic Defense Initiative: Perception vs Reality," *Physics Today* **38**, (6) 34 (1985).

Pirsig, Robert M. (1974, 1999). *Zen and the Art of Motorcycle Maintenance: An Inquiry into Values* (William Morrow & Co.).

Phua, K.K., Kwek, L.C., Chang, N.P., & Chan, A.H. (2014). *Proceedings of the Conference in Honour of the 90th Birthday of Freeman Dyson* (World Scientific, Singapore).

Rhodes, Richard (2007). *Arsenals of Folly: The Making of the Nuclear Arms Race* (Alfred A. Knopf).

Rhodes, Richard (1995). *Dark Sun: The Making of the Hydrogen Bomb* (Simon & Schuster, New York).

Russell, Bertrand (1957). *Why I Am Not a Christian and Other Essays on Religion and Related Subjects* (Simon and Schuster, New York).

Schewe, Phillip F. (2013). *Maverick Genius: The Pioneering Odyssey of Freeman Dyson* (Thomas Dunne Books).

Schlipp, Paul A., Ed. (1949, 1970). *Albert Einstein: Philosopher-Scientists* (p. 240), edited by Paul Arthur Schlipp (MJF Books).

Schweber, Silvan S. (1994). *QED and the Men Who Made It: Dyson, Feynman, Schwinger, and Tomonaga* (Princeton University Press).

Sellar, W.C. and Yeatman, R.J. (2010). *1066 and All That: A Memorable History of England* (Methuen Humour Classics).

Silver, Lee M. (1998). *Remaking Eden: How Genetic Engineering and Cloning Will Transform the American Family* (Ecco Books).

Steinbeck (1939, 1957). *The Grapes of Wrath* (Viking Press).

Tannenwald, Nina (2007). "The Threat of Weapons in Space" *Radiations* (Spring 2007), pp. 6–11. See also the introductory editorial, "Weapons in Space: Should Anyone Care?", p. 5.

Templeton Foundation Capabilities Report (2006), p.15.

Templeton, J. (1997). *How Large is God?* (Templeton Foundation Press).

Thoreau, Henry David (1960). *Walden* (Houghton Mifflin). Originally published 1854.

Tippett, Krista (2010). *Einstein's God: Conversations about Science & the Human Spirit* (Penguin Books).

Turkle, Sherry (2011). *Alone Together: Why We Expect More from Technology and Less from Each Other* (Basic Books).

Weinberg, Steven (1977, 1988). *The First Three Minutes: A Modern View of the Origin of the Universe* (Basic Books).

Weinberg, Steven (1992). *Dreams of a Final Theory* (REF).

Weiner, T. (2005). "Air Force Seeks Bush's Approval for Space Weapons Programs", *The New York Times,* 18 May 2005.

Wiener, N. (1950, 1954). *The Human Use of Human Beings: Cybernetics and Society* (Da Capo Press).

White, Theodore H. (1978). *In Search of History: A Personal Adventure* (Harper & Row).

Wiener, T. (2005). "Air Force Seeks Bush's Approval for Space Weapons Programs", *New York Times,* May 18, 2005. http://www.nytimes.com/2005/05/18/business/18spece.html

Will, George (2005). "The Oddness of Everything" *Newsweek,* May 23, 2005, p. 84.

Wren, Melvin C. (1979). *The Course of Russian History* (Macmillan).

Yates, Brianna (2016). "The 2 Windows Project," *The Echo* (SNU student newspaper) **88** (1), p. 6.

Index

www.ingramcontent.com/pod-product-compliance
Lightning Source LLC
Chambersburg PA
CBHW061235220326
41599CB00028B/5427